D0502369

James T. Campbell
Middle Passages: African American Journeys to Africa, 1787–2005

François Furstenberg
In the Name of the Father: Washington's Legacy, Slavery, and the Making of a Nation

Karl Jacoby
Shadows at Dawn: A Borderlands Massacre and the Violence of History

Julie Greene
The Canal Builders: Making America's Empire at the Panama Canal

G. Calvin Mackenzie and Robert Weisbrot
The Liberal Hour: Washington and the Politics of Change in the 1960s

Michael Willrich
Pox: An American History

James R. Barrett
The Irish Way: Becoming American in the Multiethnic City

Stephen Kantrowitz
More Than Freedom: Fighting for Black Citizenship in a White Republic, 1829–1889

Frederick E. Hoxie
This Indian Country: American Indian Activists and the Place They Made

Ernest Freeberg
The Age of Edison: Electric Light and the Invention of Modern America

THE AGE OF EDISON

Also by Ernest Freeberg

The Education of Laura Bridgman

Democracy's Prisoner

THE AGE OF EDISON

Electric Light and the Invention of Modern America

ERNEST FREEBERG

THE PENGUIN PRESS
New York
2013

THE PENGUIN PRESS
Published by the Penguin Group
Penguin Group (USA) Inc., 375 Hudson Street,
New York, New York 10014, USA

USA · Canada · UK · Ireland · Australia
New Zealand · India · South Africa · China

Penguin Books Ltd, Registered Offices:
80 Strand, London WC2R 0RL, England
For more information about the Penguin Group visit penguin.com

Illustration credits appear on pages 343–44.

Library of Congress Cataloging-in-Publication Data

Freeberg, Ernest.
The age of Edison : electric light and the invention of modern America /
Ernest Freeberg.
pages cm. — (Penguin history of American life)
Includes bibliographical references and index.
ISBN 978-1-59420-426-5 (hardback)
1. Technological innovations—United States—History. 2. Technological
innovations—Social aspects—United States—History. 3. Electric lighting—
United States—History. 4. Edison, Thomas A. (Thomas Alva), 1847–1931.
5. Edison, Thomas A. (Thomas Alva), 1847–1931—Contemporaries. I. Title.
T173.4.F74 2013
303.48'3097309034—dc23
2012039513

Printed in the United States of America
1 3 5 7 9 10 8 6 4 2

DESIGNED BY AMANDA DEWEY

To my parents

Contents

THE AGE OF EDISON

Introduction

Inventing Edison

Oil lamps burned late into the night at Edison's laboratory in Menlo Park, New Jersey, all through the fall of 1879. Months earlier the famed inventor had sent stock markets reeling with his announcement that he had solved one of the great technological puzzles of the age, the secret to transforming electricity into light. Edison promised the world that he would soon unveil a lamp that would make candles, kerosene, and coal gas obsolete. With remarkable intuition and tremendous persistence, the thirty-two-year-old inventor had surmounted one technological challenge after another in his relentless pursuit of a viable incandescent electric light. Through it all he was praised by newspaper reporters as a "wizard," pressured by his financial backers for moving too slowly, and scorned by scientific experts on both sides of the Atlantic for promising what he could not possibly deliver.

Edison felt sure that he stood on the brink of success, but he still lacked an essential ingredient: a working filament. He needed to find a substance that could handle the tremendous heat of an electrical current and, when placed in the protective atmosphere of a vacuum bulb, would *incandesce*, glowing instead of burning up. Mulling over the problem late one October evening, the inventor absent-mindedly rolled in his fingers a thread of lampblack, a form of carbon soot he had been gathering for an entirely different project. Acting on a hunch that the black sliver in his hand might provide the solution, he arranged for a test that very evening. Sealed in a vacuum, then fed current from a battery, the filament glowed long enough to convince him that he was "on the right track." Over the next couple of weeks Edison and his team tested a range of other carbon filaments—shavings of cardboard and paper, cotton soaked in tar, and even fishing line. Some failed immediately, others glowed for a time, but none worked as well as a simple filament of carbonized cotton thread. First lit after midnight on October 22, the bulb glowed through the night. Edison and his team stayed up to watch, and when the bulb finally burned out nearly fourteen hours later, they felt certain that the fundamental problem had been solved. While much remained to be done, Edison had the proof he needed that his system would work. Filing for a patent the next month, he made ready for his first public demonstration of his new light, while newspapers spread the news that the great inventor had found "SUCCESS IN A COTTON THREAD."

For more than a century Americans have regarded the creation of the incandescent light as the greatest act of invention in the nation's history, and the light bulb has become our very symbol of a great idea. We associate the bulb with a "eureka" moment, the modern version of an ancient metaphor linking light with insight. A recent study found that just the presence of a lit incandescent bulb helped a group of research subjects to think more creatively, solving problems faster than others who worked under an equally bright but less inspiring fluorescent light. As the study concludes, "exposure to an illuminating lightbulb primes bright ideas."[1]

Just as automatically, we think of Thomas Edison as the man to thank for our electric light, the one whose own burst of inspiration gave us the invention that has remained one of the signal achievements of our technological age. For good reason we remember him as one of the greatest inventors of all time, and with due respect to the phonograph and the moving picture, most consider the incandescent light his greatest legacy. For decades after Edison's breakthrough at Menlo Park, journalists often called all electric lamps "Edison's light," and long after he retired from active research in the field, the public honored him as "the Man Who Lighted the World." As one popular textbook sums it up, "A whole way of life had been revolutionized by one man's skill, insight and enterprise."[2]

Though they acknowledge Edison's great accomplishment, historians of technology have long shown the limitations of this view, which is in fact more hero worship than history. They remind us that Edison's success with a carbon filament on that October evening in 1879 was an important step, but only one of many needed to turn the incandescent light from an idea into a viable technology. Edison's achievement is clarified, not diminished, when we remember that he drew on the successes and instructive failures of many other inventors, working over decades and on both sides of the Atlantic, as well as a talented team of assistants in his Menlo Park laboratory. A closer look at how Edison, his partners, and his rivals developed the first working electric lights shows that the inventor does not pull insights from a void, like a bulb suddenly illuminating a dark room, but that invention is a complex social process.[3]

Tracing the early decades of electric light, we see that Edison relied on a transatlantic exchange of science and technology. While acknowledging a debt to European science, many nineteenth-century Americans took patriotic pride in their country's rising international reputation as a leader in technological innovation, and considered this talent for creating new machines to be a vital expression of their democracy. Declaring the United States *a nation of inventors*, many opinion leaders credited their society's remarkable mechanical ingenuity to its broad public education system, its liberal patent laws, and its egalitarian faith in the practical genius of ordinary men and women. These seemed like the essential ingredients for a culture vibrant enough to invent an Edison.

The evolution of electric lighting in its first decades also reminds us that Edison and his rivals raced not to perfect a science experiment but a commercial product, a lamp that would not only work but would *sell*, and in sufficient quantity to reward a huge investment of capital. As Edison's biographer Paul Israel puts it, "While Edison the individual is celebrated as the inventor of the electric light, it was the less visible corporate organization of laboratory and business enterprise that enabled him to succeed." This heavy capital investment, funding a new approach to product research and development, not only enabled Edison to create the first viable incandescent lighting system, but also played an essential role as the inventor's first fragile lamps evolved into the twentieth-century electrical grid.[4]

When we recognize that Edison was not a lone genius but an important member of a much broader transatlantic culture of invention, we better understand the origins of the unsettling and exhilarating wave of technological creativity that we have been riding ever since. Men and women of the nineteenth century were the first to live in a world shaped by *perpetual* invention. Fascinated by the new machines that were transforming their lives, the public eagerly toured exhibitions of the latest technologies and followed news about inventions in the popular press. Many Americans tried their own hand at invention, dreaming of winning a patent or two of their very own. We see in their infatuation with new machines the birth of a paradox now familiar to all who live in a modern industrial economy—the unsettling sense that change is the only constant, that the steady arrival of marvelous but often disruptive inventions has become a matter of routine.

As each new invention arrives we take the old ones for granted. To modern readers, electric light has become so pervasive that its remarkable qualities are buried under a thick layer of the obvious. And so it is easy to forget the excitement and wonder that Americans felt when they saw it for the first time, their giddy sense that they were crossing a threshold into our modern age. In every city, town, and hamlet across the country, one evening marked the moment when electric light arrived, and citizens always turned out to give the technology a warm welcome, honoring the historic event with music, speeches, parades, and the ritualistic burial of their old oil lamps.

While electricity marked just one more milestone in a push for stronger light that had been transforming urban life for at least two centuries, the new lamps created a change in kind. Electricity produced a cleaner, brighter, and safer light than oil or gas could provide. As importantly, incandescent bulbs proved infinitely adaptable, offering a much more complex and subtle mastery of light. Electric lamps erased the ancient obstacles posed by open flames, shedding light where candles, oil lamps, and gas jets were of little use—from the deep recesses of the human body to the bottom of the ocean, from hairpins to uncharted polar regions, from baseball fields to battlefields. As people worked out the implications of electric light's flexibility and power over the decades, they created a world not just brighter but *illuminated,* its public and private spaces carefully engineered to produce what one lighting specialist called "the moods and illusions" of modern life.[5]

In this sense electric lighting has become a tool of social control, a device powerful enough to induce in modern crowds a range of feelings, from greed to euphoria to reverence. Still, people were never simply passive consumers of the new light but played an active role in its creation. In various forums, from sermons to science fiction to oil paintings, the whole culture grappled with its meaning, while many used their own powers of invention to adapt the technology to a range of new uses that no single inventor could have anticipated. When Edison demonstrated his first working light bulbs at Menlo Park in the last days of 1879, this marked the culmination of a long and complex process of invention. But it also staked out a much broader field of technological creativity as many inventors, most long forgotten, worked to realize the light's enormous potential. As the technology spread in the late nineteenth century, all who saw the light knew that it was good—but it took decades more to realize all that it was good for.

Some who helped to invent the electric light never set foot in a laboratory and understood little of the strange new language of electricity. The lighting system that emerged by the 1930s was shaped not just by inventors, scientists, and industrialists, but also by insurance inspectors and progressive economists, by lamp designers and store window trimmers, by health reformers and union workers, and by specialists in countless other fields who adapted the light for their own purposes. Indeed, it would be hard to find a line of work in early-twentieth-century America that did not consider the new light a valuable tool. Photographers and theater artists embraced its possibilities, as did hunters and fishermen, policemen and educators. The light transformed the fields of urban planning, architecture, and interior design and contributed to a late-nineteenth-century revolution in the fields of surgery and public sanitation. The technology also played a fundamental role in the era's Industrial Revolution. Within a generation, electrical manufacturing became one of America's largest industries, while improved lighting created a safer, more efficient, but intensified production process in many other fields, from coal mining to cotton spinning.

Others adapted the light bulb to build a richer urban nightlife. While social critics cursed the new technology for encouraging a shallow but seductive mass culture of flicker and flash, many more found the creative display of electric light to be beautiful, exciting, and fun. Colorful electric signs, theater marquees, spotlighted window displays, and even Christmas lights—all were turn-of-the-century inventions that have come to epitomize modern ideas about the good life, and all were made possible by the inventive use of incandescence.

As Americans worked to realize all of electric light's possibilities, many also saw that the light was creating them—changing their relationship to the natural world, shaping the rhythm of their days, and transforming their culture. This new regime of intensified light energized some and exhausted others. Doctors warned that electricity's light disrupted sleep patterns, creating a new generation of frenetic but feeble and nearsighted Americans. Others welcomed what they considered to be humanity's ultimate victory over the dark. Edison himself claimed that electric light was "improving" human nature, and many agreed that strong, clean light made cities safer and workers more productive and happy, while encouraging a nightlife of stimulating sociability that country folk could only read about under their smoky kerosene lamps.

In short, the electric light intensified the pace of city life in the late nineteenth century, contributed to the era's rapid expansion of industrial production and consumer culture, and helped create urban America's new mass market for entertainment. In the process, the new light stimulated countless innovations, new machines and new ways of living, which were greeted by men and women who were both eager and ambivalent—just as we are today when we encounter the latest new technology. By tracing the role that electric light played in the pivotal decades when our modern culture was born, we can better appreciate that inventions are not simply conjured up by great men like Edison but evolve as they are shaped by a variety of political, economic, and cultural forces.

Those forces continue to shape technology down to our own day, and in the lighting field the pace of invention has quickened now that concerns about climate change have forced us to acknowledge what Edison realized long ago—that incandescent light is an inefficient technology that wastes the world's limited energy resources. Even as Edison worked to sell his lighting system around the world, he joined the quest for a better alternative, and in his own lifetime the electrical industry's push for greater energy efficiency produced improvements in every aspect of his basic design. If Edison could see us now, he might well be surprised at the long run that his incandescent lamp has enjoyed. But it also seems likely that he would be an eager participant in our current search for a better light bulb.

One

Inventing Electric Light

In the fall of 1881, vast crowds elbowed their way into the grand salon of the Palace of Industry in Paris, host to the world's first International Exposition of Electricity. They lined up for a chance to "ride electrically" in a railcar and to listen to the strains of opera on the telephone, transmitted from a nearby theater. Others strolled through a model Parisian apartment of the future, each room a showcase of electrical ingenuity. In the kitchen, stoves churned out "electric waffles"; a living room, warmed by electrical fire, rang with the music of an electric piano; the bedroom featured electric hairbrushes—simply turn a switch and "offer your head to the gentle caresses." Even play could be improved with electrical power, as the nursery of the future displayed electric trains and dolls, and a working toy telegraph. Back in the main hall, hundreds of fingers pressed hundreds of buttons, provoking a chaotic din of electric gongs and buzzers, while beneath it all thrummed the bass notes of the dynamos that powered the proceedings.

The exposition gathered the greatest minds in the new science and showcased novel applications of electricity from inventors on both sides of the Atlantic. While these pioneers shared their research and judged the relative merits of the new machines, most who attended understood only a fraction of what they saw. One British visitor captured this very modern state of mind. The display filled spectators with "high expectations of great wonders for the future," he thought, but left them more awed than informed. "Three quarters of the public felt a sort of bewildered admiration for something wonderful they did not understand," he concluded. "The other quarter gaped as good charity school children." Offering this first peek into the world's electrical future, inventors were asking people to put their faith in forces they could not fathom.[1]

Still, none left Paris doubting that humanity hovered on the verge of a great transformation. Always feared as a mysterious and dangerous force, electricity now obeyed human command. Dynamos seemed to conjure its current out of thin air, while batteries allowed people to store power, release it at will, and even carry it around in a box. The exhibits marked no limit to the uses for this "tireless energy of the universe," no field of work or play that would not soon be improved by the modern world's alliance of scientists, inventors, and industrialists.

The latest and most wondrous of these inventions was the electric light. True, many visitors had seen some form of electric light before; the last great Paris exposition in 1878 featured a dazzling display that reinforced the French capital's much older reputation as the "City of Light." But progress in electrical lighting had exploded since then, and the 1881 exhibition provided the world's first grand showcase of all that had been accomplished in just a few years. For the first time, dozens of new lighting systems would blaze side by side, competing for top honors and lucrative contracts. Jointly, they made the case that electricity was "the light of the future." Most exciting of all, Thomas Edison used this opportunity to show-

The exhibit hall of the International Exposition of Electricity, Paris, 1881.

case the incandescent lighting system he had unveiled for the first time only months before, a feat that commanded attention around the world. Plenty still doubted Edison's claim that he had solved the technical challenges of incandescent lighting, but all wanted the opportunity to see for themselves.

Visitors to the 1881 Paris exposition would never forget the sight. Together, Edison and his rival inventors installed a "terrific mélange" of more than twenty-five hundred lights. "Strong, fresh and dazzling," the lamps filled the vast hall with a "gorgeous flood of brilliant yet mellow radiance." No one had ever seen anything like it, and Americans in that crowd felt it had been well worth crossing the Atlantic to see. The lights did much more than cast a festive gleam on the exposition's "carnival of human skill." Many of the new electrical apparatus on display in Paris—the electric brushes, toy telegraphs, and self-playing pianos—were curiosities and conveniences. But a machine that could create enough cheap and powerful light to hold the night at bay was a breakthrough of a different magnitude, liberation from one of the fundamental and primordial limits imposed by Nature on the human will. The lights of Paris seemed a beacon pointing the way for all of humanity "toward a deeper wisdom, a freer, broader and purer existence."[2]

A century and more after the late nineteenth-century explosion of technological creativity, we are the jaded beneficiaries of all the machines we have inherited from our inventive Victorian ancestors. We know more than they did about technology's power to destroy as well as to liberate, and about the environmental costs that come with ever more powerful machines. On the other end of a century of mass production, we also see more clearly that human desires are insatiable, and that one of the defining characteristics of a modern industrial economy is the manufacture of artificial longing.[3]

But these twenty-first-century doubts may prevent us from seeing the electric light as it appeared to those who flocked to the Paris exposition more than a century ago. A light that could burn without spark and smoke, and promised to turn vast swaths of night into day, was rightly hailed as a "marvel" and a milestone in human history. And yet it was only one of many transformative inventions produced in this era, a time when miraculous feats of human ingenuity became almost routine. As one observer at the Paris exposition summed it up, its displays contained "wonders enough to have made our grandfathers doubt whether the reign of law in the order of the universe had not come to an end; but these miracles of ingenuity are accepted by the men of this generation as a matter of course."[4]

While this technological revolution was transatlantic in origin and scope, many European visitors conceded that the United States had become "the wonderland of modern civilization," a fount of new machines, each embraced by American society with astonishing speed. "With a vigor and determination for which we Europeans have not example," a German visitor wrote, the Americans had "learned how to harness steamships, railways, the telegraph system, and agricultural machines to their uses." An Englishman complained that "Europe teems with the material product of American genius." The United States had become "the leading nation in all matters of material invention and construction, and no other nation rivals or approaches her."[5]

Americans had spent much of their first century grappling with a sense of cultural inferiority to Europeans. But over the course of the nineteenth century they grew justifiably proud of the lead role they had assumed in the more "practical" field of technological invention. "So many inventions have been added," as Ralph Waldo Emerson put it in 1857, "that life seems made over new." Americans understood that they lived in a thrilling time, an age of remarkable technological feats, and they welcomed these with a mixture of awe and nervous excitement, patriotic fervor, and even religious enthusiasm.[6]

Of course, men and women who lived during the dawn of the electrical age were not naïve and uncritical enthusiasts for the inventions that were transforming their lives. They experienced the painful birth of a new industrial order, as the massive new corporations that shaped this process brought with them troubling disparities of wealth, political corruption, labor strife, and environmental degradation. In what Mark Twain dubbed this "Gilded Age," many sensitive critics cast doubt on their culture's growing faith in technological progress, complaining that the new machines trampled on "imagination and poetry" and seduced people into mistaking material wealth and power for true "human peace and happiness." The great prophet of Concord had his doubts about where all this was leading, warning that the new tools of modern industry had some "questionable properties." "We must look deeper for our salvation," Emerson advised, "than to steam, photographs, balloons and astronomy."[7]

Many more, however, scoffed at this sort of talk. The new technologies looked to them like "benevolent agencies," capable of giving men and women more control over their lives than any generation had ever known—power not just for the wealthy few, but available in some measure to all. The hunger for more light was not a yearning induced by clever promoters; the dark was, for all of human existence, a palpable and universal obstacle to human happiness. Throughout history, to be in darkness was to be diminished, shuttered from the world, and in the late nineteenth century the yearning for more light seemed more urgent than ever. Many forms of work in the new industrial age, both in factories and offices, made more demands on the eyes, requiring greater attention to detail. At the same time, the urban world had only grown darker, as tall buildings cast their shade and the unregulated burning of coal belched a smothering pall on urban areas, blocking sunlight and coating every window with grime. More than one commentator declared that "more light" was the crying need of the modern age. A machine that could conjure light at will, then, was an evident good. "Comply with electricity's conditions," one Gilded Age writer rhapsodized, "then but turn a key, and the servant of all life will be present in light and power."[8]

Sensing that their age had scored an unprecedented victory over the dark, many writers enjoyed telling the story of history's ascent toward electric light, a tour through the pine knot torch, the tallow candle, the whale oil lamp, and most recently the invention of gaslight in the early nineteenth century. Gas itself had been an exciting and controversial technological marvel only for a few decades, the first public gas streetlamp having been erected in Baltimore in 1817. That electricity now threatened gas with an early extinction confirmed that history had entered a new phase, an era of perpetual technological progress in which marvels of the future would be arriving faster than ever. To trace the evolution of lighting devices in this way was to describe the growth of civilization itself, many thought—a story that placed the incandescent bulb at the summit of an ancient quest to conquer both mental and physical darkness.[9]

The public had been fascinated by electricity for more than a century before inventors created the first lighting systems, when Enlightenment thinkers first began to study the various manifestations of electricity and identify them as facets of a single elemental force. These early "electricians" built simple machines to generate static electricity, and learned to store the mysterious electrical "fluid" in glass Leyden jars. The line between groundbreaking science and parlor trick was often blurred in these years, and some savants quite literally shocked their audiences and made their hair stand on end. The study of electricity leaped forward in 1800 when the Italian physicist Alessandro Volta developed the first battery, creating and storing electricity through the chemical interaction of zinc and silver discs. While Europeans did most of this early scientific research, Americans pointed with pride to their own Benjamin Franklin, whose investigations into lightning were universally admired.[10]

The great pioneer in the use of electricity as a light source was the British scientist Sir Humphry Davy, who followed Volta's lead into what he called "a country unexplored." Early in the nineteenth century he demonstrated not one but two ways to turn electrical energy into light, marking out the field of future research for de-

TOP, *Sir Humphry Davy demonstrates the first electric lights at the Royal Society, London, circa 1810;* BOTTOM, *visitors examine the massive voltaic battery Davy used to produce sufficient current.*

cades. He showed an audience at the Royal Society the brilliant light produced when an electrical current passes across a narrow gap between two carbon rods; as the current leaps the gap, the tip of each rod burns white-hot, and remains so as long as a proper distance between those rods is maintained. The four-inch bend of light between the carbons formed an arch, a term later shortened to "arc." In those same years Davy also illustrated the principle of incandescence, producing light by passing an electrical current through a platinum wire, making it hot enough to glow. Here was the first ancestor of Edison's lamp, and every incandescent bulb ever since.[11]

Davy made no attempt to translate these experiments into a functional source of light—it was only a by-product of his broader interest in what would become the field of electrochemistry. The prospects for producing a viable commercial lamp were remote anyway; to power his own lighting experiments, he relied on a cumbersome array of 220 linked battery cells, the largest source of electrical power in the world at that time.[12]

But over the ensuing decades inventors on both sides of the Atlantic pursued the idea, most often devoting their efforts to the more powerful arc. Some of these men were master mechanics or electrical technicians who learned their trade by working on the telegraph. Some were solitary tinkerers who had to cajole skeptical reporters to their chambers to show off their creations, while others shone their experiments from a rooftop or in an urban plaza. One New York inventor demonstrated his light in an oyster house, then ran his wires out the window, lighting the city's first electric streetlamp on Greenwich Street.[13]

Many of these early lights drew public excitement, some drew considerable financial backing, but none came close to delivering a commercially viable rival to gas, kerosene, and whale oil. The odds for success were much improved when both the British scientist Michael Faraday and the American Joseph Henry discovered the principle of generating electricity through induction, relying on magnetism rather than the chemical action of batteries—a breakthrough that by the 1870s led to the creation of increasingly powerful dynamos, machines that produced more and steadier current at a much lower cost. The British employed arcs on lighthouses, and led the way in adapting the new technology to their navy ships as a weapon of war. When a powerful searchlight could not "actually put out the eyes of our enemies," one British electrician explained, it could "at least deprive them of the opportunity of carrying out their designs unobserved in the dark." Parisians used the arc light in a handful of factories and railroad stations, as well as some theaters and elegant department stores. Some Americans saw their first electric light at the nation's great Centennial Fair in Philadelphia in 1876, but most paid more attention to displays of the massive Corliss steam engine, Edison's multiplex telegraph that sent several messages on a single wire, and another remarkable electrical device, Bell's new telephone. The principle of the arc light had been known for a half century, one American journalist observed that same year, but "till the present time it can hardly be said to have come into practical use."[14]

This was because the arc light was a marvel with problems. Its light was not only overwhelming, but it also flickered, creating a disconcerting strobe effect. A reporter who witnessed an early demonstration in New York described these "trying fluctuations of brilliancy" this way: "For a few seconds the light would be strong, clear, unwavering; then it would flicker, and darkness for an instant would fall. . . . Suddenly the light would grow brighter, sometimes increasing slowly in brilliancy, at other times leaping at a bound to the full measure of radiance." The light thrilled, but pained.[15]

The early arcs also demanded constant attention, as the carbon rods or wicks burned rapidly, requiring a skilled technician to replace them every few hours. A wayward wick not only created those annoying flickers but sometimes shed shards of glowing ash that trickled down on anyone below. The only thing more bothersome than the lamp in action was its opposite. If the gap between carbons grew too large, the circuit was broken, plunging an urban crowd into a darkness more sudden and disorienting than any that Nature had ever produced.

Many inventors tackled these glitches, and in the late 1870s one after another announced that he had developed "THE ONE THING NEEDFUL." A dozen different designs arrived on the market, offering unique ways to solve the arc light's well-known problems. When Parisians lit some grand boulevards for an international exhibition in 1878, they installed arc-light "candles" designed by the Russian refugee Pavel Jablochkoff. These represented the state of the art at the time, and for many the streets of Paris provided the first demonstration of electricity's potential to transform the urban night, improving the city's capacity for business and pleasure "beyond all estimate." Reports filtered back to America about "this grand fire," one that made gaslights look "yellow, muddy and petty" in comparison. Jablochkoff had turned a few blocks of Paris into "the most magnificent spot in the world at that moment."[16]

The gas men enjoyed a quick revenge, however, on those who were too eager to toss them into the dustbin of history. Within months after the 1878 exhibition, the Municipal Council of Paris abandoned plans to make the Jablochkoff candles a permanent fixture of the City of Light. Like so many designs before, the electric candles proved a disappointment—expensive, complicated to maintain, too often flickering or going out altogether. Observing these developments, *Scientific American* decided that "up to this stage of the contest the victory rests with gas."[17]

Many of these nagging problems were effectively solved by an American inventor, Charles Brush of Cleveland. While the arc light emerged out of the laboratory experiments of a British scientist and made its grand public entrance on the boulevards of Paris with help from a Russian inventor, Charles Brush did what Americans were increasingly known to do—he took a conceptual breakthrough from Europe and with clever and persistent mechanical tinkering found ways to make the device simpler, more reliable, and cheaper to run. Before long many Europeans were buying their arc lights from Brush, and the American lights spread to markets around the world.

Charles Brush grew up on a farm in northern Ohio, but from the start showed far more interest in science than farming. Poring over the popular scientific magazines available even to a farm boy in mid-nineteenth-century America, he found an "endless source of delight" in the latest news about breakthroughs in astronomy, biology, and physics. When not doing farm chores, he made his own batteries, electromagnets, and other simple electrical devices. When he succeeded in assembling enough batteries to power his own crude version of an arc light, the experience filled him with "joy unspeakable."

Recognizing their son's unusual aptitude, his parents sent him to a technical high school in nearby Cleveland. There he embraced the stimulation of a city that was fast becoming a center of industry and technological innovation. Before heading off to the University of Michigan to study engineering, Brush capped his high school career with a commencement oration on the "Conservation of Force." In it he urged his listeners to contemplate the enormous power pent up in a lump of coal; a chunk of ancient sunlight it was, only waiting for the right invention to liberate its hidden potential, transforming it into a new form of light for the modern age.

The dynamo was just the machine needed to liberate the light trapped inside coal. Inspired by reports about electromagnetic dynamos developed by European inventors, Brush designed and constructed one of his own in his family's barn, powering it with a small engine turned by a team of horses. In a remarkable year of "eager experiment," he continued to improve his design and in 1877 won top honors in a competition sponsored by Philadelphia's prestigious Franklin Institute. In a fever of creative genius, Brush simultaneously developed a new arc light, creating an electromagnetic device to control the carbon rods that was simpler but more efficient and reliable than any that had come before.

Brush's new light would soon shine in city squares, rail stations, and palaces around the globe, but its debut on the streets of Cleveland lacked any sense of grandeur. His workshop overlooked a major boulevard, and one evening as a parade passed under his windows he fired up his new lamp. He heard startled cries on the street below as the procession was stopped dead in its tracks by "the brightest light they had ever seen." Brush expected a welcoming party to rush upstairs to congratulate him on his accomplishment, but instead he faced an angry policeman who ordered him to "turn out that damn light!"[18]

Undaunted by this unenthusiastic early review, Brush arranged with the city fathers to give his invention a proper unveiling in downtown Monument Park on a Saturday night in April 1879. This time Cleveland was ready to give the twenty-nine-year-old inventor his due. When Brush flipped the switch on twelve arc lights, filling the square with instantaneous illumination, a crowd of ten thousand roared its approval, while ships moored in Lake Erie fired cannon volleys and a band let loose with "brassy strains of triumph." Those assembled gaped at the familiar landscape with fresh eyes, seeing every vivid detail of the surrounding buildings. Even the statue of Commodore Perry, at the far edge of the park, jumped out in "bold relief," the wrinkles on his bronzed brow clearer than day, and the crook in his finger somehow more expressive than ever.

For those who had not been in the square that night, reporters groped for words to describe this new, unearthly light, comparing it to "the brightest moonlight that the romantic mind could conceive of." Using a test that others would apply and refine in the years

Brush's arc lamp produced a brilliant light by passing a strong electric current across a narrow gap between two carbon rods.

ahead, one reporter explained that "reading was a matter of perfect ease." All agreed that the gaslights that still burned in the park looked "really dyspeptic" by comparison. Cleveland's city fathers rewarded Brush with a contract to light the downtown square for a year, twelve lights for a dollar an hour—the first blow in what would become a decades-long struggle between gas and electric companies for control of the streets.[19]

Soon after, Brush moved his light show to the much smaller town of Wabash, Indiana, where he proposed to light the entire town with a single cluster of arcs mounted in the cupola of the new city hall. The news spread for weeks, and thousands came on special excursion trains. They pressed together in the dark streets as Brush started his dynamo, filling the town with what one observer called a "strange weird light, exceeded in power only by the sun, yet mild as moonlight." Unlike those in Cleveland, the folks in Wabash greeted the sight not with huzzahs and cannon blasts, but with silence. As one reported, "The people, almost with bated breath, stood overwhelmed with awe, as if in the presence of the supernatural." Unprepared for the sight, a farmer on the outskirts of town felt sure that the Second Coming had arrived. For months afterward, passing trains halted in their tracks so passengers could stare at the light.[20]

Innocent of some of the basic laws of thermodynamics, many who peered at Brush's new dynamo believed they were seeing a machine that manufactured energy—and would soon flood the world with a limitless and practically free source of light. Instead, Brush took a considerable loss on his early installations. But he wisely reckoned that once people became accustomed to walking nighttime streets lit bright as day, they would never want to return to the comparative gloom of the gaslight.

Ever after, the citizens of Wabash have proudly pointed out that theirs was the first city to light its streets entirely by electricity—though this accomplishment marks both their progressive spirit and the small size of their town, compact enough to be lit with a single arc-light installation. At first, not many other towns were ready to cancel their gas contracts and take a bet on the new technology, so Brush and his competitors looked for customers elsewhere. Most Americans encountered their first electric light not in the exhibition halls of a scientific society or a grand public building like the Paris Opera House, but in places like the front of a doctor's office in Cincinnati, Brush's first private contract. In Boston he showed his light at a mechanic's fair, then arranged a temporary installation in front of a department store. New Yorkers' first glimpse of Brush's light came when he set up a display on the roof of a Jersey City railroad depot, casting a beam across the Hudson River strong enough to catch the eye of newspaper reporters in the Tribune Building. They too noted with satisfaction that, though miles away, Brush's light was bright enough to read a newspaper by.

On the Fourth of July, 1879, Brush turned the full force of his sixteen lamps on the thundering flood of Niagara Falls, showing that he could produce "electric rainbows to order in the darkest night." To his chagrin, one paper mistakenly credited Thomas Edison with the display, confused by Edison's widely publicized announcements about his progress on an incandescent light. "It is a little tough," a Brush company spokesman complained, "that when Edison hasn't a light in operation outside his laboratory, he should get credit for the work of Mr. Brush, who has over 500 in use."[21]

Brush found early customers among department stores, most notably Wanamaker's, and in the lobbies of elegant hotels like Chicago's Palmer House. He sold lights to a Coney Island amusement park, billiard parlors and bowling alleys in New York's Bowery district, and factories making everything from thread to iron. His sixteen-light systems were not only powerful but also portable, and soon became a fixture for traveling theater troupes, itinerant revivalists, and the circus. When journalists reported on the arrival of the circus in their town in these years, they often skimmed through their accounts of daredevil acts and exotic animals, saving their breathless enthusiasm for the new electric light. See the circus, they urged, but *be sure to go to the evening show.*[22]

As Chicago customers entered the grand tent of Bailey's circus on a spring evening in 1879, they gaped at a half dozen balls of smokeless and unconsumed fire burning over their heads, spraying "a million little needles of variegated light." Some came prepared, and stared through smoked glass at the fearsome spots, like "molten metal." Others looked no farther than their own hands, which looked somehow more visible than ever, as clear and crisp as an "illustrated picture." Skipping the show within, a crowd gathered outside the tent where two men tended the dynamo that fed current to the arc lights. The tenders warned men with pocket watches to stand their distance, as the dynamo would magnetize the timepieces and render them useless. They also encouraged some to touch a copper plate near the dynamo, and enjoyed a laugh when these unsuspecting electro-tourists recoiled from a powerful shock. The machine produced so much power but looked so simple—just copper wire whirring around a magnetized iron bar—and yet the electricians boasted that a single Brush light, raised high enough and properly reflected, would turn the darkest night into day for a space of fifty miles. A tall tale, but in an age of such wonders how could a journalist be expected to sort grandiose lies from science's new truths?[23]

The advocates of arc light declared that gaslight looked "simply ridiculous" when compared to its modern rival, but they conceded that many still felt a "prejudice" against the new light's cold, harsh shine. The Scottish author Robert Louis Stevenson was among those who waxed nostalgic for gaslight long before it ever disappeared. Its warm flicker had turned the nineteenth century into a "new age . . . for sociality and corporate pleasure-seeking," he thought, providing a stimulating release from the "murk and glimmer" of times past. But the display of electricity's "ugly glare" on the streets of Paris served as fair warning that the "old mild lustre" of gas would soon be eclipsed by "a new sort of urban star," one which "shines out nightly, horribly, unearthly, obnoxious to the

Brush arc light over New York's Madison Square, 1882.

....

human eye; a lamp for a nightmare! Such a light as this should shine only on murders and public crime, or along the corridors of lunatic asylums, a horror to heighten horror."[24]

Stevenson was not the only one who dreaded an electrified future that looked as ugly as it was inevitable. But many seemed to take comfort in widely circulating newspaper stories about how less sophisticated people around the globe were reacting to the new light. For example, an electrician carrying the first arc on a steamship up the Missouri River could not resist seeing how a group of Indians gathered at the shore would react when he cast the beams of this "white man's light" on them. "The astonished aborigines were paralyzed for a moment," he reported with sadistic glee, "and then they set up a dismal chant, lay down and rolled over and pawed up the sage brush, and made the ambient air tremble with their antics. Finally assured that the big medicine of the white man was harmless, they assumed an attitude of quiescent bewilderment." Back east, many Americans were not quite sure themselves whether they should clap or cringe when they saw their first arc light, but such stories about "terror-stricken" and "superstitious natives" at least gave them some assurance that the new light was "theirs," and put them on the leading edge of civilization's advance.[25]

Set up in Madison Square and other urban parks, the arc cast an alluring false moonlight that drew hundreds of city dwellers each evening—but some found the light so harsh that they shielded themselves from it with umbrellas. One observed that "people look ghastly—like so many ghosts flitting about." Health experts warned that too much exposure to the new light would cause eye diseases, nervous exhaustion, and freckles. In order to minimize these complaints, arc-light companies began to encase their flaming wicks in thick globes of opal glass. While this produced a mellower effect, taxpayers complained about the paradox of paying for an expensive new light, only to tamp it down by more than 50 percent in order to use it. Gas men particularly enjoyed shaking their heads over this evident waste of coal.[26]

When the arc light came indoors, illuminating large interiors such as railway stations, exposition halls, and ballrooms, the public reaction was also mixed. As the lamps consumed carbon, they emitted an unpleasant hum that some compared to the sound of swarming bees. More damning, the harsh light bleached out colors, gave food an unappetizing hue, and cast an unforgiving spotlight on the human face, exposing every wrinkle, blemish, and stray hair. Even one of the light's defenders had to admit that its blue tinge did not "show off to advantage the natural beauty of the Anglo-Saxon race." Another declared the light a boon to brunettes, since it was "death to the blond. The pinkest of them take on little shadows under the eyes, and purple tints come into their lips and their cheeks get ashen. The effect upon the artificial bleacher is simply cadaverous." After a bad experience under the arc, some women vowed never to be seen near electric light again.[27]

So the arc light worked fine in a rail depot, city plaza, or circus tent—or if you needed to light up Niagara Falls, or blind your enemies in a naval battle. But bringing the electric light into homes demanded a different approach, and long before Edison took up the challenge, many recognized that incandescence might provide the solution. In the decades following Humphry Davy's experiments, most inventive energy had gone into arc lighting, but a smaller number of inventors had tried their hand at the technical challenges of creating a commercially viable incandescent light.[28]

As early as the 1840s, J. W. Starr, a young inventor from Cincinnati, developed a working bulb using both carbon and platinum filaments. Starr took out an English patent, and demonstrated his light in the United States and Britain, putting twenty-six bulbs in a candelabrum, one for every state then in the Union. Working in the era before dynamos, however, he was forced to rely on expensive battery cells for his power, and was said to have literally "worried himself to death" over the problem. Another inventor, Heinrich Goebel, a German immigrant living in New York, created working incandescent lamps twenty years before Edison "without suspecting that there could ever be a world market for them." Moses Farmer, a prolific New England inventor, rigged a set of small battery-powered incandescent lamps that could run for an entire month. The lights amazed his neighbors in Salem, Massachusetts. Once Farmer realized that his invention consumed battery power at a rate far more expensive than gaslight, he went on to devote his energy to other inventions. In 1872, a Russian inventor, Alexander Lodyguine, revived interest in incandescence, demonstrating before an assembly of European scientific dignitaries a lamp that sent current through a strip of charcoal sealed in a glass tube. Joseph Swan, a chemist and largely self-educated inventor from Newcastle upon Tyne, England, had been trying on and off for almost two decades to create a light bulb. His experiments led him, as early as 1860, to create incandescent bulbs using carbonized paper filaments; but, lacking the ability to create more than a partial vacuum, his lights burned out quickly. Distracted by other projects and the pressures of family life, he set his experiments aside.[29]

Thanks to successes and failures of these early efforts, the nature of the problem was well understood by the time Edison took up the challenge in 1877. He joined a crowded field that was already looking for a steady and inexpensive power supply, a method for "subdividing" that current so that multiple lamps could be turned on and off at will, and a filament sealed in glass, capable of reaching terrific temperatures, glowing without burning up.

The new dynamos being developed in the 1870s solved the problem of supply, replacing the much more expensive batteries in use up to that point. Edison's interest in joining the race for a viable incandescent light was inspired by his encounter with an improved dynamo recently constructed by the Yankee inventors Moses Farmer and William Wallace. Not long after Edison purchased one of these, reports circulated that he was on the trail of incandescent light, and would soon unveil a "scientific marvel." Arc lights already proved that electricity could produce plenty of light, but Edison announced that he had solved a more difficult "puzzler": how to divide this force into many small lights. "I have it now," he boasted. Up till that point his rivals had all been working along the same "groove," he explained, but he would astonish them soon with the simplicity of his solution.[30]

As Edison studied his dynamo and began to sort through the many technical challenges of replacing gas with electricity, he came to envision not just a new type of incandescent lamp, but a central station capable of distributing light and power over many city blocks. He planned to invent a lighting system, not just a bulb. His improved dynamos would send current through buried copper wire mains and into each house the same way gas traveled—but Edison felt sure that he could do so at a "trifling cost" compared to that rival. "I can light the entire lower part of New York city," he predicted, "using a 500 horse power engine." The elaborate system of gas lines need not go to waste, he boasted to reporters. Once the gas companies closed down, their pipes would serve as ideal conduits for the new electric wires—a suggestion that offered cold comfort to the gas men.[31]

But Edison had declared "mission accomplished" too soon—even in the flush of his early declaration of victory in 1878, he conceded that he still had not resolved the problem of the bulb. He needed something that would cast a pleasant light, and not "waste away" when exposed to heat. He conjectured that platinum might do the trick, but he treated the whole matter as a detail that could be resolved in short order. "Now that I have a machine to make the electricity I can experiment as much as I please," he told reporters, as if that settled the matter. In the end, it did.[32]

When one reporter asked Edison about the vast fortunes to be made in patenting the first commercially viable incandescent light, he insisted that he cared little for money—what motivated him was the love of "getting ahead of the other fellows." Edison was not the only inventor with this sort of competitive streak, and not the only one driving himself to find a solution before others beat him to it. Some observers were ready to bet instead on Jim Fuller, a reclusive New York mechanic-inventor who was rumored to be well ahead of Edison. Fuller pursued a different approach to incandescent lighting, and to protect his secret he worked as much as possible in seclusion. But he worked too hard, ruining his health as he drove himself to perfect his creation. Just as he was putting the finishing touches on his latest lamp design, he collapsed. He called an assistant to his side, reviewed the intricate technical details of his invention, and then reviewed them once more. After his friend assured him that he understood all that he had heard, Fuller "leaned back in his chair, with a look of satisfaction in his face, and instantly died." Fuller expired with a smile on his lips, knowing that his invention would survive and that his patents were secure. They soon proved worthless.[33]

Edison had that same drive, but he also enjoyed significant advantages over Fuller and "the other fellows." His previous track record of success, first with telegraph improvements and then with the phonograph, gave the young man a mythic status as the "Wizard of Menlo Park," and the capital he needed to create his invention laboratory at Menlo Park. There he assembled the latest equipment, a vast range of raw materials, and a staff of assistants who brought technical skills or scientific training that Edison lacked. In this Edison invented a new style of invention, a coordinated program of scientific research and product development that amplified the speed and range of his individual genius by channeling it through the talents and insights of dozens of assistants.[34]

While none doubted Edison's track record, his genius, and his resources, rivals in the race for a working incandescent light searched the news stories about Edison's great breakthrough and found no evidence that he had actually discovered anything new. The American Hiram S. Maxim, for example, considered himself ahead of Edison, and was surprised to see the man praised so lavishly in the newspapers not for "what he had done, but what he was going to do." As Edison tackled the incandescent light problem, he carefully studied the strengths and weaknesses of the approaches already tried by his competitors, and decided the key to his success would involve a filament made of platinum. Others had already tried platinum but had abandoned it in favor of carbon, an element trickier to handle but inexpensive, and capable of withstanding the high temperatures of an electric current when placed in a vacuum. In short, when Edison first announced that he had solved the great problem of a viable bulb, his rivals had reason to believe that he offered "absolutely no novelty."[35]

Still, the great man's boasts threw the London and New York market for gas stocks into a "perfect panic." The consensus of scientific opinion notwithstanding, not many investors wanted to bet against Edison and his Menlo Park invention factory. On the strength of his hunch, a consortium of financiers and lawyers organized the Edison Electric Light Company, and put a hundred thousand dollars at his disposal to press forward with his experiments. All those funds were needed, and many more, as the Menlo Park team soon came to realize the numerous technical challenges they still faced.[36]

Though newspaper readers and stock investors followed the news of Edison's progress closely, few in the public could follow the technical issues. Nor did they understand much about how invention worked, the actual process of discovery taking place not only in Menlo Park but in workshops on two continents—the dozens if not hundreds of skilled men involved; the endless calculations, false starts, and painstaking adjustments; the uneasy marriage of theoretical knowledge and manual dexterity; and the high-stakes economic struggle to win the race for the essential patents. Instead, for most Americans invention seemed to be embodied in a single self-taught young man from Ohio who seemed endowed with almost godlike talents of intuition and persistence.

The legend is nicely captured in the Edison jokes that became a staple in the newspapers. Once Edison's lamp succeeds, one asked, "What is to be done with the sun?" Others announced that Edison had dropped the lamp altogether to pursue new brainstorms—a chilled cow that produced ice cream, or a "365 shirt," a garment made of that many thin layers of paper. Each morning for a year, the user only had to tear off the previous day's layer and go about his business. A "366 shirt" was in the works for leap years. Another paper announced that the government had granted several thousand patents in the previous quarter—only a couple went to someone besides Edison, and those two were thought to be worthless. Many of these tall tales came from Edison himself, who beguiled reporters with his folksy charm while he and his Menlo Park team worked desperate hours to make good on his premature claim to have solved the challenges of creating a reliable incandescent light.[37]

Edison produced his first working lamp a few months later, one that used a filament of platinum wire and an ingenious device for regulating its temperature so it would not be burned up by the current. When Edison made a small demonstration of this lamp to a friendly reporter, those who understood the technical issues, including Charles Brush and distinguished scientists in America and England, declared the invention impractical. It was a "philosophical toy," they said, too expensive to produce, too delicate to maintain, and too inefficient in its consumption of electricity. Newspapers reported that the staff at Menlo Park was in despair because "Edison claimed to be able to do what he could not."[38]

Though stung by what he considered to be unfair slanders against his new light, Edison ultimately had to agree with his critics. While he had once boasted, "I alone stand on the platinum metallic method," he now abandoned this approach; the platinum lamp, he conceded, belonged in the "cemetery of inventions." He joined his rivals in a search for a suitable form of carbon filament, while offering no further public demonstrations of his progress. Some, impatient over this detour, began to ridicule him as a "newspaper electrician" who had gulled the world into buying his extravagant boast that he would soon offer "the brilliant illumination of the earth at a cost not worth mentioning."[39]

This roller coaster took another sharp turn when Edison announced, in late 1879, that he would throw open his Menlo Park laboratory for a public demonstration of his new lighting system. Encouraged by their limited success with carbon filaments of lampblack and then cotton thread, Edison and his team had spent long hours in the laboratory, looking for the best source of carbon. After trying everything from raw silk to cork, horsehair to beard hair, they settled for the moment on filaments made from loops of carbonized cardboard. These still fragile prototypes could burn up to three hundred hours, each casting about the same light as a large gas burner. On New Year's Eve, trainloads of sightseers arrived at Menlo Park to witness "Edison's latest marvel," his laboratory and grounds glowing with the light of about fifty of these incandescent bulbs. One Edison biographer described the event as a "spontaneous mass festival," as thousands pressed in to turn the light on and off themselves and to shake the great inventor's hand.[40]

Journalists witnessing the early display tried to convey this "surprising experience" to their readers, the first of many writers who searched for the words to describe the look and feel of incandescent light. While Edison's system represented the very cutting edge of technological progress in 1880, witnesses were struck less by the new light's utility than by its surprising organic beauty. The bulb was "a little globe of sunshine," one wrote, which shed a "bright, beautiful light, like the mellow glow of an Italian sunset." Beneath an enormous layer of human contrivance, after all, the light was still a glowing ember of carbon, and in that sense not so different from any lamp that had come before.[41]

But visitors felt sure that Edison's bulb was now "the highest form of illumination known," a leap ahead of its rivals, the gaslight and the arc. "It is the flash of a thousand diamond facets perpetuated," one explained. And yet, unlike the arc, it was "in no sense dazzling to the eye. . . . You can stare at it and feel no weariness of vision." The advantages over gas were duly noted, as they would be many thousand times over for another half century to come. "There is no flicker," one reporter noted. "There is nothing between it and darkness. It consumes no air and, of course, does not vitiate any. It has no odor or color." For all of human history, illumination had been a multisensory affair—something not only seen but felt and smelled. But Edison ushered in a new age when the lamp provided "simply light."[42]

While reporters who witnessed the New Year's Eve display rhapsodized, drawing thousands more to Menlo Park over the next ten days, some who had not seen the new lamps with their own eyes remained skeptical. For decades, others had produced incandescent lights, many using a similar approach—carbon inside a globe as free of oxygen as possible. Those others had proven both expensive and fragile, so why should Edison's be any different? Suspecting that Edison was more showman than scientist, one French electrician ridiculed his latest invention as "the new electric playthings of a semi-practical prestidigitator. Amusing, perhaps . . . but a complete failure." British papers, proud of the lead role of their nation's scientists in the field of electrical research, reminded readers that while Edison was "one of the most ingenious of men," he was more a canny salesman than a "scientific expert, judged by the English standard." Just weeks after the Menlo Park demonstration, another electrical expert claimed the entire thing had been an illusion—a few dozen lamps might burn long enough to impress reporters, but the fundamental challenges had not been solved. "Every claim [Edison] makes," he insisted, "has been tested and proved impracticable." "The promises have been very profuse," another noted, "while the performance has been practically nothing."[43]

Such objections only grew louder after the Menlo Park demonstration, as Edison once again closed off his laboratory from public scrutiny and went to work trying to solve the many technical problems that remained before he could bring his system to market. He still needed to work out the complicated challenges of safely burying his wires and delivering current at a reasonable cost across a wide area, with switches that allowed customers to turn their lamps on and off without disrupting the entire system. The bulb also needed further improvement, leading Edison to conduct a worldwide search for the best source of carbon. In the end he settled on filaments made of thin carbonized slivers of Japanese bamboo. Meanwhile, his corporate backers raised millions on the strength of his brief Menlo Park demonstration, while gas stocks once again veered wildly. Far from hailing Edison as an inventor hero, many now accused him of either deluding himself or, worse, playing on his

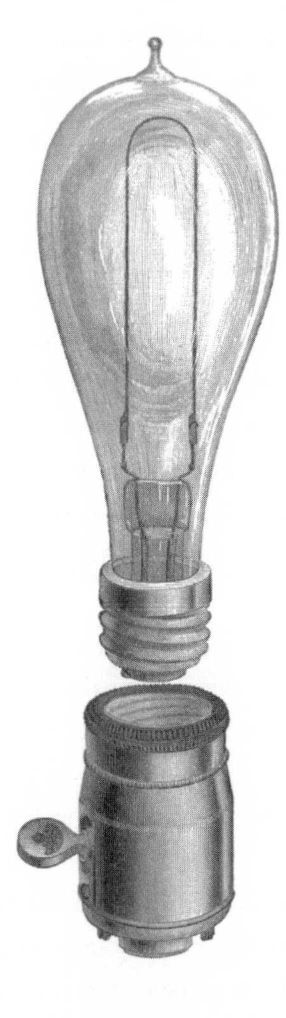

An early Edison bulb. The bent filament of carbon was held in place with platinum wires and sealed inside a vacuum.

reputation as part of a "base and unscrupulous manipulation of stocks," cashing in on what one called his "electric bamboozler."[44]

Leaders in the gas industry eagerly joined this chorus of Edison's doubters. The inventor had "nothing new" to offer the world, they assured their own investors, just one more quixotic attempt to burn a carbon filament in a vacuum globe. This form of electric light was a "well known fact," decades old by 1879, but no one had come close to creating a system that worked on a scale, or at a price, that threatened the gas companies. Accusing Edison of participating in a

Wall Street stock fraud, a gas company spokesman observed with much satisfaction that since Edison's demonstration at Menlo Park, gas stocks had recovered nicely.[45]

As months passed, one reporter noted that all of the lights Edison had put on display on New Year's Eve had since burned out. Menlo Park's houses had gone back to burning kerosene, while the town's only streetlights were the moon and stars. This writer concluded that it was difficult, "amid the fierce disputes of the electricians over Mr. Edison's work, for an unscientific and unprejudiced on-looker to clearly see just what the inventor has accomplished." Others also sensed a backlash against Edison, not just from jealous Europeans or his economic rivals, but even from his many American admirers who had only recently been "blindly hailing him as the high priest of electrical knowledge." Edison protested that he had never claimed that his lamps were indestructible, and that his team needed more time to perfect his invention. But the longer he delayed, the more he gave the impression to many that "there is a loose screw here."[46]

Even those who did not doubt his eventual success raised serious questions about his originality. As far as they could tell, Edison's approach offered nothing substantially different from lamps offered by his rivals, several of whom had already demonstrated working incandescent bulbs, and held the patents to prove it. Encouraged by improvements in the vacuum pump, and a growing public interest in the pursuit of a viable lamp, Joseph Swan in Newcastle resumed his decades-old experiments with a carbon-filament bulb. Swan demonstrated his incandescent bulb in early 1879 and took out a British patent in November 1880. Repeating the performance that year, Swan tactfully noted that Edison's first lamp, using platinum wire, had "not realized the hopes of the inventor." While Edison had been pursuing that dead end, Swan finally succeeded with his own carbon filament, placed in a hermetically

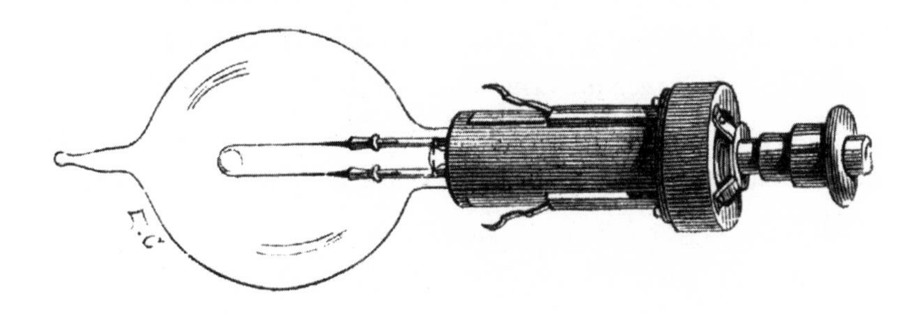

Joseph Swan's incandescent bulb, 1881. Swan demonstrated a working carbon-filament lamp in January 1879 and won some key British patents before Edison.

sealed bulb, and demonstrated his system five weeks before Edison's Menlo Park event. With characteristic modesty, Swan added, "I do not mention these things in any way to disparage Mr. Edison, for no one can esteem more highly his inventive genius than I do. I merely state these facts because I think it is right to do so in my own interest, and in the interest of true history." Swan installed his system in Cragside, the home of a scientifically adventurous patron, and in the spring of 1881 set out the world's first incandescent streetlamp in front of his shop, drawing several thousand curious spectators. "For hours the crowd stood and gazed," one paper noted. "The illuminations were so attractive that people were evidently most reluctant to leave them."[47]

Another English inventor was not far behind. In early 1881 English papers reported that St. George Lane Fox had exhibited his own version of the incandescent bulb, which he had been working on for two years, at a private exhibition attended by a half dozen dukes, marquis, and colonels, who declared the lamps a "perfect success." Where Edison used cardboard and Swan used cotton thread, Lane Fox found success with "dog's tooth grass . . . impregnated with oxide of zinc." Miffed at all the attention Edison received for his Menlo Park demonstration, Lane Fox pointed out that his own patent for a working incandescent bulb preceded Edison's. While insisting that he was "the last man to depreciate Edison's ingenuity," he added that "nobody could claim exclusive rights in incandescent lighting."[48]

Closer to home, Edison's American competitors included William Sawyer, a New York inventor who, in spite of limited funds and a drinking problem, managed to develop a working carbon lamp by the spring of 1878, using a carbonized paper filament sealed in glass. Sawyer charged Edison with patent infringement, publicly declared him a fraud, and may even have come to Menlo Park to attempt industrial sabotage. In turn, Edison dismissed him as a "despicable puppy." That seems a fair judgment, as Sawyer shot and killed a man after an argument over Edison's light, and died of alcoholism while awaiting his trial.[49]

Not long after Edison opened his laboratory for public inspection, another American inventor, Hiram Maxim, brought to the New York market his own, small-scale incandescent system, one that Edison insisted had been plagiarized. Edison even tried to block Maxim from showing his bulbs in Paris, but Maxim resisted when French authorities tried to seize his lamps. The French left the matter to be settled in one of many long court battles that loomed ahead.[50]

The Paris exposition of 1881, then, served as a most visible place for an excited public to see for themselves how well the incandescent light could deliver on its brilliant promises. Newspapers on both sides of the Atlantic followed each turn in the drama that pitted the gas companies against the underdog inventors. Were they intrepid visionaries, the vanguard of a new electrical age, or were they false prophets?

Would incandescent light triumph over gas for interior lighting? Could this invention explode one of the world's most heavily capitalized and politically influential industries? Could Edison's plan for distributing central power really deliver enough energy, and at a price that the powerful gas companies could not undercut?

Even if the technology worked as its backers predicted, how many customers would be willing to let this mysterious and deadly force into their homes and businesses?

And if electric light managed to fulfill all of its grand promises, providing clean, strong, and safe light in almost limitless supply, which system was better, and which inventor's lifetime of work would soon be relegated to the attic of technological curiosities?

These were the questions posed by the Paris exposition, and debated in papers and scientific journals everywhere in 1881. The controversy had only grown more interesting as so many rival inventors staked their claims within months of each other. In this debate, few who pondered the future of the new technology were in any position to make a meaningful prediction. On both sides of the Atlantic, reasonable-sounding voices weighed in, offering wildly varying conclusions. Some heard the death knell of gaslight and welcomed the dawn of a limitless and free supply of power. Others forecast that electricity was a passing fad, trumped by self-interested inventors like Edison and his Wall Street backers.

Edison came well prepared to this showdown in Paris, which was billed as the first "open and equal competition" between rival electrical systems. As one proud American journalist described it, the inventor shipped to France a display designed to "stun foreigners," including dozens of detailed, working exhibits of his groundbreaking contributions to the telegraph, the phonograph, and the telephone. But the centerpiece of the Edison exhibit was a comprehensive display of his new incandescent lighting system—the first chance for the public to see it outside of the brief demonstration at Menlo Park. Edison's display dwarfed all others, and was powered by the largest dynamo ever made, the "great machine from America" that was soon nicknamed "Jumbo." Defying the calculations of many of Edison's critics, this 220-ton dynamo was both larger and far more efficient than anyone else had thought possible.[51]

Edison's company displayed its lights on the exhibition hall's grand staircase, where visitors encountered two massive electrified *E*'s, and a huge portrait of the inventor that revolved under a spotlight. And so the first person to see his name "up in lights" was Edison, because he put it there himself. Such bombastic displays of inventor worship were no reflection of the inventor's ego, but rather of his company's canny instinct for self-promotion, a talent fully encouraged by his corporate backers, who recognized that their "wizard's" legendary reputation would serve them well in the pending struggle over patent rights and in the worldwide drive to win customers.[52]

As visitors entered the grand hall, so many lights blazed from so many different systems that the effect was magical, but not helpful for those eager to make an informed choice about the relative merits of the rival systems. This was easier in the smaller galleries that allowed each inventor to showcase his lamp's peculiar excellence. Even then, judging the quality of each inventor's light was tricky business. Every system performed better on some days than others, and they all suffered embarrassing breakdowns at some point during the fourteen-week exposition. Some praised the Edison lamp, which now used a filament of bamboo fiber, for the "remarkable uniformity of its texture and light-giving power," while others found Swan's lamps more appealing—though his mixture of stronger- and weaker-powered bulbs got mixed reviews. Some found the Swan lamps brighter than Edison's, while others praised them for being softer—evidence of either the variable performance of these new technologies or the subjective nature of human perception of light, or both.

Whether or not Edison's lamps outshone his rivals, they were just the most visible part of a remarkably complex and beautifully elaborated system. Edison won the exhibition's only Gold Medal of Honor for electric lighting. One distinguished English expert, William Preece, reported to his fellow electricians back in London that it was time to take back the "many unkind things" they had been saying about Edison. Preece admitted that he had once expressed grave doubts, not only about Edison's lamps but also about his character. In the midst of an unrelated dispute over a microphone patent, he had called Edison a "Professor of Duplicity" who had "a vacuum where his conscience ought to be." But now he had to concede that the American had "at last solved the problem that he set himself to solve." Privately, Edison enjoyed hearing that he'd made Preece "eat boiled crow."[53]

Edison could make no claim that he had invented the first working light bulb—the patent offices and newspapers provided ample evidence that others had accomplished this feat months and even years before. What he had done was create a complete lighting system that linked his powerful and more efficient dynamo, through a central main, feeders, and switches, to an incandescent bulb of superior design. His system delivered a steady supply of current to hundreds of lights, at varying distances from the source of power, and used parallel circuits to maintain the current even when some of his lights burned out or were turned off. Alone among these rivals, his bulbs used a filament of high resistance, a crucial innovation that saved money by using a relatively small amount of current for each lamp. In Paris he showed that electric light not only worked but could be distributed some distance from a central station, a system with the potential to become large and economical enough to challenge the gas companies. Other inventors in Paris had shown that they could light a house; Edison was well on his way to lighting an entire city neighborhood.[54]

Some left Paris entirely converted, sure that they had seen "the light of the future." Good riddance, they said, to "greasy candles, filthy oil lamps, and poor, unhealthy gas." In the new age of incandescence, lamps would provide light and nothing more—not the smoke, soot, and heat of gas and oil, nor the sputtering buzz of the arc lamp. Surveying the rapid progress of electrical ingenuity on display in Paris, one reporter predicted that many of his readers would "live to wonder how they could have endured any other light." Another thought that their children would find it hard to understand why they had ever tolerated gas, "a subtle and unmanageable agent, deleterious, dangerous, and nauseous." The answer had to be an innate human "fascination in brilliancy" that led people to put up with gas even when they knew it was harmful to their health. "They are like moths," he concluded, "content to die in flame."[55]

But others had reason to doubt that electricity would soon win an easy victory over gas. For one, nothing in the exhibition offered much to calm public fears about the potential dangers of electricity. Protected by only the flimsiest of insulation, and in some cases none, the electric wires set as many as five different fires during the Paris exposition. One man almost ignited himself by leaning over a balcony railing, accidentally allowing his watch chain to cross several wires—a problem he only noticed when his chain burned bright red and nearly set fire to his waistcoat. Joseph Swan fretted that his own system might cause a fire in the plaster and lath building, and spent part of his time in Paris pacing and muttering that he was "cutting too close to the wind." Looking back years later, his assistant concluded that "a special providence looks after children." Burning down the exhibition hall, he speculated, "would have damaged our future business outlook most seriously."[56]

Others knew that the gas companies would do everything in their power to snuff out their new challenger. From the start, all recognized electric light as an agent of creative destruction that would only survive and thrive by stealing away gas customers, and in this titanic struggle no one could think of the powerful gas interests as the underdog. They were among the most heavily capitalized companies in the Western world, their interests well protected by a cozy relationship with city governments everywhere. Even in Paris, the early showcase of electric light, the gas company maintained a powerful hold on the city government's lighting contracts. And when Jablochkoff tried to establish a beachhead for electric light in England, bringing his candles to the Thames embankment, London's gas men had done a good job of discouraging the invasion of their turf. When Joseph Swan drew crowds in Newcastle upon Tyne by erecting an incandescent streetlight, the gas company had countered by waging a "battle of the lamps," erecting a powerful set of gas burners across the street and declaring "a victory for the old over the new." All understood that the same sort of contest would be fought everywhere people already used gas to light their streets, homes, and factories.[57]

People could see with their own eyes that the electric light was preferable to gas, but could only accept as a matter of faith Edison's claim that his system would be able to deliver illumination over a large territory and at a reasonable rate. As one visitor to Paris summed it up, the future seemed to belong to electricity, but "as yet the matter of expense stands obstinately in the way." That was Edison's next, and perhaps greatest, challenge: to translate this great act of technological genius into a functioning system that could be imposed on the physical, economic, and political reality of city streets, businesses, and private homes—and survive a rough-and-tumble struggle with the mighty gas companies.[58]

An early Edison dynamo, more efficient than others thought possible.

....

Two

Civic Light

While American visitors to the 1881 Paris exposition marveled over the display of lights in the hall, those who wandered a few blocks away were surprised to find that electric lights had not spread far in the City of Light. One reporter from St. Louis noted that he had seen many more arc lights on the streets of his own hometown. In fact, American cities and even smaller towns embraced the new technology with a speed and enthusiasm that Europeans soon found both fascinating and reckless. While the United States was no leader in the science of electricity, inventors such as Brush and Edison had not only developed the most effective working systems but backed them up with an entrepreneurial initiative, even aggressiveness, that was fast making America the world leader in the commercial development and installation of electric light. Electric lighting systems, first arc and then incandescent, became a booming business, a popular enthusiasm, and a reform crusade that swept the country in the 1880s. This movement effectively served both mammon and morality. The wide-open field promised nice profits for entrepreneurs while also pleasing those who considered electric light a tool of civic improvement and progressive reform, capable of making life more comfortable for the crowded city's middle-class residents, and improving both the health and the virtue of its toiling masses. In one town and city after another, politicians, business leaders, and editors exclaimed, "THE ELECTRIC LIGHT—WE MUST HAVE IT!"

While Cleveland, Wabash, and a few other towns made early experiments with arc lighting in public streets and parks, in most places private investors took the first risks with the new technology. Factory owners usually led the way, since they were willing to pay extra for a light that produced more efficient workers and fewer fires and explosions. Large retailers also tried lighting systems, using the technology to lure curious customers. Even the earliest of these arc-lighting systems provided more light than most store or factory owners could use, and so they often rented their surplus lamps to adjoining stores, or to the city, which paid a dollar or more per night for each light, enough to illuminate an adjoining street or two. In this way many towns and cities eased their way into the electrical future, investing no significant capital on the untried experiment. Before long, however, they were forced to respond to a growing "agitation" for more light in public spaces, a yearning to extend the day that grew in strength now that a solution seemed at hand.[1]

The greatest American showcase for electric street lighting was a three-quarter-mile stretch of Broadway that came to be known as the original "Great White Way." In 1880, Charles Brush installed twenty-three arc lights along this central artery of American commerce, from Union Square to Madison Square. Just after dusk on a December evening, reporters and city officials gathered in the company's new power station to watch the treasurer's young daughter give a ceremonial flip of the first switch. At the last minute, though, her father could not bear the risk of her electrocution, and the honor passed to the steam engineer. As he turned a small lever all of the powerful lights came on simultaneously, "like stars emerging from the darkness." Startled Christmas shoppers turned away from store windows and looked out across the scene, marveling at familiar sights suddenly cast in a new light. A *New York Times* reporter delighted in the "artistic effects" of intense brightness and deep shadow, while conceding that "unaccustomed gazers" might have found the whole scene too brilliant and painful to bear. "The great white outlines of the marble stores," he reported, "the mazes of wire overhead, the throng of moving vehicles, were all brought out with an accuracy and exactness that left little to be desired." While the city always turned off its gaslights around midnight, the carbon arcs flamed on until sunrise, "lighting up the deserted streets with unwonted splendor."[2]

Leaders in other cities rushed to keep up, sending delegations to see Broadway for themselves and to investigate the claims of the various lighting systems. Just months after his early demonstrations, Charles Brush expanded his company dramatically, but still struggled to keep up with demand. The light's first entrance into each new town was always grand, a cause for civic celebration. At the moment when Quincy, Illinois, fired up its first fourteen lamps, for example, the mayor arranged for the old Civil War cannon to be fired. The whole town celebrated with a band concert, while the hometown editor proudly declared Quincy the best-lighted town in the West, with the possible exception of Dubuque.

As this sort of talk suggests, the market for electric light grew in part because Americans embraced the idea that their town's standing on the great ladder of civilization could be measured by its ability to provide residents with the latest technological conveniences. Each time one town or city unveiled the light, boosters in neighboring municipalities felt the sting of inferiority and fretted that their town might be doomed to bring up the rear of history's

In 1880, Charles Brush turned a stretch of Broadway into the nation's first "Great White Way."

march. In Los Angeles, for example, city leaders had been following reports of the early tests of street lighting with a mixture of scientific curiosity and civic envy. The *Los Angeles Times* pouted in 1882 that "many of the Eastern cities, of the same size and much less importance than Los Angeles, are now lighted by electricity." Even worse, the city's close rival, San Jose, had pulled ahead, placing arc lights on a two-hundred-foot-tall iron scaffold that loomed like an Eiffel Tower over the town's main intersection. Spurred by an appeal from the editor of the *San Jose Mercury*, hundreds of citizens had donated funds to erect the tower, and soon boasted that they lived in the only town west of the Rockies that had "risen to the dignity of being illuminated by electric light." Reporting on San Jose's accomplishment, the *Times* editor made no attempt to hide his covetousness. "THE ELECTRIC LIGHT," he wrote, "*Los Angeles Wants and Must Have One.*"[3]

Within months the Brush Company answered the call, erecting a half dozen arcs on tall towers. The results proved so satisfactory that the city abandoned all public gas lighting, proudly asserting its claim to be the first city in America to do so. "The people of Los Angeles feel rich in having the light," the *Times* beamed, especially when laggard towns like Oakland and San Francisco came south seeking advice about the new technology. Though eastern cities might consider Southern California "remote from the center of civilization," residents could hold their heads high, assured that few places in America were "further advanced in the conveniences of the present age than our own Los Angeles."[4]

In Atlanta, Henry Grady and his fellow New South boosters followed the arrival of electric light in rival cities across the South, and fretted that their hometown had gained a reputation as "the poorest lighted city of her size in the country." All that changed when the first arc-light company arrived in 1883. For months, as the contract was hammered out and subscribers signed on, the *Atlanta Constitution* reported each step toward the city's brighter future. Even the arrival of a shipment of light poles merited a report, and brought a small crowd of the curious down to the rail depot. "Let Us Have Light!" the paper urged, and as the sun set on a December evening, residents flocked downtown to see the "great treat." "The bands of darkness will be broken," the newspaper had predicted that morning, "and a flood tide of beautiful white light will be emitted from the handsome brass lamps now being distributed over the city."

The electricians fired up only three lamps that night, thanks to a special arrangement with P. H. Snook's department store, a downtown emporium that used this chance to attract a horde of Christmas shoppers. Sadly for Snook, while he enjoyed a bump in holiday sales, his store burned to the ground a month later—an accident attributed to a fault in the new wiring. Despite this embarrassing setback, more lights went up over the next few weeks. Months later, hundreds of Atlantans still gathered on hot summer nights to swim in the pools of street light. "A well lighted city is a pride to any section," Henry Grady boasted, and the new light was essential equipment for "a city with the enterprise and needs of Atlanta."[5]

In 1891, a Frenchman touring the United States visited a tiny hamlet on the Oklahoma prairie that proudly lit several arc lamps each evening, "useless" light that shone down on the town's empty main street. "There is no need for them whatever," the visitor observed, but concluded that settlers for miles around, isolated in their primitive farmhouses, looked toward those lights each night and became more "confident of the future of Oklahoma." Touring the tiny towns of the far West, he found that an American "tells whether his small town is flourishing or not by three tests—electric light, water works, and street-cars. These form the three ends of his ambition, and you will never make him confess that the city of his adoption is not one of the wonders of the world, once these three services are efficiently organized."[6]

Many nineteenth-century commentators believed that their own age was characterized by a particular hunger for light. Perhaps that's because, in the densest parts of urban America, the world had grown darker. Coal consumption skyrocketed in the rapidly growing cities and industrial valleys, providing heat to residents and driving the steam engines of an increasingly mechanized economy. City air filled with soot and smoke, and on bad days produced "fogs" dense enough to turn midday into an eerie dusk. The electric light, first on streets and soon in homes and offices, seemed just the thing to pierce through this man-made murk, the only way to compensate for the loss of sunlight in the modern industrial city. No one noted the irony that coal-powered electricity produced both smoke and light, the poison and the antidote—most likely because the only option less polluting than electric power would have been to live with less light.[7]

Gilded Age cities seemed not only darker but more dangerous, and lighting companies marketed their product as nothing less than a police force on a pole. After nightfall, urban parks became notorious danger zones, a haven for the city's dregs and an infernal playground of indecency. Now all that could end, not by converting sinners or reforming criminals, but by harnessing light's power of exposure. As the mayor of Baltimore put it, "An electric light is a nocturnal joy to an honest man, but a scarecrow to a thief." Friends of the light in Los Angeles put the formula more dramatically. "The brighter the light," they reasoned, "the better for truth, purity and honor, and the worse for fraud and all that fearful spawn of evil which flourishes in the darkness." Frustrated by the slow development of electric street lighting in England, British reformers applauded what they called this "American theory" that "each electric light is as good as a policeman."[8]

Throughout the 1880s, as cities erected powerful arc lights in city parks and boulevards, they hoped to win control of these civic spaces for law-abiding citizens. Most working people toiled until well after dark, especially in the winter months, so strong lighting helped to make these places useful to more people, more of the time. For example, those advocating for the "respectable working girls" on New York's East Side called on the city to install lights in the riverfront park to stop "roughs" from using the cover of darkness to insult women. The chairman of the Republican state committee in Illinois even found the light useful in cleaning up Chicago's notori-

"The Electric Light in Its Moral and Social Aspect,"
Electrical Review, *1885.*

ously corrupt elections. He spent thousands of dollars on election night in 1886, rigging powerful locomotive headlights at the polling stations to expose the Democratic "ballot-box stuffers, shoulder hitters and ruffians" who usually controlled the polls after dark. As soon as the lights came on, he reported with great satisfaction, "the scoundrels could be seen slinking away into the alleys and shadowy places." Temperance and vice reformers also embraced "the modern light" as a tool for exposing their neighbors' "deeds of darkness." Under the cover of night, otherwise respectable men felt no shame when they tumbled through the streets at all hours, and had to be "taken home by the police." When lechers, thieves, and drunks were forced to do their evil deeds under strong streetlights, the eyes of every person in town were drafted to serve in a voluntary police force.[9]

Storeowners adopted similar strategies to protect their premises at night. Though some balked at the expense of burning a light long after the store had closed, others advised that a well-placed electric light did double service as a burglar alarm. One sales manual urged retailers also to install a large clock in the center of their store, somewhere visible from the street. That way, passersby who wondered about the time would peer through their store windows, thus keeping an eye on the premises all through the night.

In its heyday, gaslight had once been sold in just the same way, but the friends of electric light pointed out that robbers could easily turn off the gas, and that gaslight was only half as powerful as electric, and thus only half as effective in fighting crime. In fact, some found gaslight congenial for prostitution, and pulp novelists used the idea of "gaslight" to suggest the opposite of safe streets—its flickering yellow flame now seemed lurid, the perfect light for a debauch or a murder. New York electricians argued that the city's police records confirmed this argument, as arrests for robberies declined steadily each year in the decade after electric lights were installed. Such claims are suspect, of course, as the stronger light more likely had the same effect on criminals that it does on cockroaches, not eliminating them but simply pushing them into darker corners of the city.[10]

The link between strong light and safe streets became so axiomatic that some worried that a prolonged blackout would produce a crime wave. At a time of intense labor unrest, fears centered on the possibility that anarchists or some other working-class mob might target gas and electric lighting plants, using the cover of darkness to wreak havoc on government and property. That threat seemed realized one night in Los Angeles when "mischievous hoodlums" managed to turn out the lights; fearing this blackout was the first step in an "organized scheme to plunder the city," the mayor called out the entire police force to handle the phantom menace of a revolutionary rabble. Though strong lighting made city streets less dangerous, many fretted that public order now depended on an unreliable new technology, one that erected only a fragile barrier against chaos and criminality.[11]

City officials liked to put electric lights in busy thoroughfares and major parks, places where they would not only be most useful but also most valuable as a sign of the city's cosmopolitanism. Merchants on the most affluent boulevards used bountiful light to lure customers, and the wealthy burned light as a sign of social status, not only in their homes but also in elite clubs and the lobbies of posh hotels. Thus strings of electricity marked the contours of some of the most affluent zones of the city, a visible marker of the line that divided the haves from the have-nots. But in their desire to use the new technology to fight crime, police chiefs pressed to install more streetlights in the poorest and most crime-ridden places. New York's police chief reported that the owner of one of the city's most notorious brothels begged him not to install lights near his business, but in vain. The chief wished he could double the amount of light thrown upon the worst streets in town, giving them what he called "the light remedy." Lowlifes thus did their beleaguered neighbors an unintended favor, hastening the introduction of lighting systems that might otherwise remain the exclusive privilege of more respectable neighborhoods.[12]

Muckraking journalists eager to alert their readers about the plight of slum dwellers invariably emphasized the fact that this "other half" not only lived in mental and moral darkness but were physically deprived of light. As Jacob Riis guided his readers into a typical tenement, he warned them not to stumble in the "utter darkness"—hallways and alleys one had to feel rather than see—until he exposed them through the lens of his camera. In the same way, other civic reformers insisted that light would not only cut down on crime in the slums, but would harness the power of exposure to promote urban renewal. Shining a strong beam on a slumlord's neglected properties would shame him into cleaning up alleys and improving his buildings, they insisted, an argument that granted electric light a power of exposure that even broad daylight could not match. In this way, those attempting to cope with the growth of urban problems in the Gilded Age added lighting systems to their tools of reform, along with better sanitation, playgrounds and parks, and public health initiatives. The windowless halls of tenements needed electric lighting, they insisted, while well-lit streets would quickly pay for themselves, raising property values as crime rates fell.[13]

Electric light companies pressed the point, even circulating testimonials from a convicted burglar who had given up his life of crime because "electric lights are death to our trade." The value of this testimonial was undermined by the fact that he offered it while on his way to prison for murdering a cashier. But even in prison, he could not escape the probing beam of electric light, since penitentiaries and asylums were among the earliest public institutions to install electric lighting. Wardens found that in addition to being safer than open flames, electric light gave them the power to impose lights-out in a cellblock with a single switch. And besides, they found that electric light made these institutions more cheerful.

While electric light made the urban night less dangerous, it also made it less private, exposing behavior that was not illegal but illicit. Electricians liked to joke that the gaslights flickering in most city parks were so dim that one needed a lantern to find them. But for young lovers, that shadowy half-light offered opportunities. Everyone understood that in a world of crowded tenements, city parks provided a place not only for "breathing" but also for courting. Some authorities bathed public spaces with light specifically to "repel lovers." As modern glare drove out the dark, one paper observed that "bench spooning" had become a lost art. Perhaps for this reason, some Yale students waged a war on the "detested light" in 1885. "Let the unromantic light be removed to more useful fields," the school paper insisted, "and let the primeval gloom be restored to the campus elms." Some students underlined that request by chopping down the campus's only light pole, and later shooting out its globe so often that the administration posted a police guard. The light company finally gave in to the "yahoos of Yale," agreeing to move the pole. Students exulted in their successful bid to restore "privacy to their spring and summer evenings."[14]

By 1885, more than six hundred lighting companies had been formed, representing more than a dozen rival arc and incandescent systems—more electric light than could be found in all European countries combined. That year downtown Chicago boasted over a thousand arc lamps, using no fewer than nine different patented systems. Towns without lights soon became more of a curiosity than those that had them. The Boston suburb of Newton, for example, held out for years; the thrifty Yankees running that town considered electric light an indulgence, and thought that all residents should be in bed by ten o'clock anyway. A decade into this process, journals such as the *Electrical World* duly reported each new contract, but found that the edge of excitement had gone. The flip of a switch on a new dynamo brought nothing more than "quiet congratulations and business-like calculations."[15]

Not so out in the western territories, where residents hungered for the new light as a proof that they were no longer on the frontier and had joined the ranks of civilization. European tourists were surprised to find that even tiny hamlets banded together to start electric light plants. The seven hundred residents of Monticello, Minnesota, burned sawdust to fire their dynamo, while other prairie towns experimented with windmills. In the Oklahoma Territory, the push for electric light began surprisingly early. Two days before the famous land rush opened the region to homesteaders eager to stake their claims, federal officers caught a band of settlers who had illegally entered the territory, hoping for a jump on the competition. When the marshals found them, the interlopers had already marked out streets and town lots and were making plans "to advertise for bids on an electric light plant." Instead they were bound as prisoners, driven to the border, and ordered to "never again set foot in Oklahoma."[16]

Residents waited more patiently in the Wyoming Territory, but when the Edison company installed Laramie's first lights in 1886 the town celebrated in style, marking the event with a fireworks display, a balloon ascension, and a band concert. Capping it off, the assembled crowd serenaded the electricians, thanking them for this "great modern blessing." The local editor exulted that "Laramie can now, without fear, invite comparison with any city west of the Missouri River." Lots of small but ambitious towns felt the same way, pointing to their new lights as proof that they "must not flippantly be called a village" anymore.[17]

Not all westerners welcomed the change. The humorist Bill Nye, for example, mourned the arrival of electric light in Wyoming as one sign of the passing of more romantic times on the old frontier. Progress had arrived, littering the trails with discarded fruit cans and filling each gulch "with the odor of codfish balls and civilization." The West had passed into history now that one could ride for a day in any direction and find oneself reading a newspaper under the electric light. Lampooning rural America's hunger for technological respectability, Nye wrote an essay comparing ancient Babylon with modern-day Cheyenne, a skewed history lesson he told from the vantage point of a small-town booster. The ancient cradle of civilization had enjoyed a three-thousand-year head start over Wyoming, he crowed. And yet, "Cheyenne has the electric light and two daily papers, while Babylon has not so much as a skating rink."[18]

When city officials or private entrepreneurs decided to purchase a lighting system, they faced a bewildering range of choices. By the mid-1880s, Brush and other arc-light entrepreneurs had added incandescent lighting to their offerings, while Edison and other incandescent lighting companies sold arc lights. A factory owner or town alderman had to sort through competing bids and rival claims from a half dozen different companies, each making impressive promises. Company representatives provided alluring brochures full of testimonials from ecstatic customers, and short lessons explaining the alien terminology of electrical power. One insider satirized the usual sales pitch by summing it up this way: "There are two kinds of electric lights, namely, our kind and the other fellow's kind. Our light is much better than the other fellow's light. The other fellow's light is surrounded by a cloud of non-luminous verbosity."[19]

Making a difficult decision even worse, customers understood that rival inventors were busy dragging each other into court over patent disputes. Bet on the wrong inventor and one might end up party to a lawsuit. And in those early years, some companies proved better at selling stock and winning contracts than they were at delivering light. In Memphis, for example, a company made a great show of putting up poles and stringing wires, but in the end local investors found that they were the ones being strung along. New York police finally caught up with another of the new industry's "errant scoundrels," a distinguished-looking German who traveled under a half dozen false names, selling dynamos he did not have in cities from Cuba to Australia. "Complaints come in daily from all parts of the universe," one paper noted.

Even when civic leaders and lighting salesmen bargained in good faith, both entered into uncharted territory. Eager for a foothold in this lucrative market, the electric companies usually offered to install the entire system, from dynamo to lamps, at their own expense. In return, they asked for the city's lighting contract for the year, typically promising to charge the same as the gas company. Gas men squawked, but the electrical entrepreneurs offered impressive statistics about the flood of light their arcs could unleash over the town. Confident in the superiority of their product, and needing much less capital than the gas companies to go into business, the electric upstarts were willing to bet that once a town's residents enjoyed the new light they would never want to go back, even if the eventual cost turned out to be more than the annual gas bill. For no additional investment, the town was promised much more light. Few could resist giving this a try.

But in these early days both the buyer and the seller of electric light hardly knew just what sort of bargain they were making. Aggressive salesmen from the lighting companies quite literally promised the moon, suggesting that they could provide enough candlepower to create the effect of a full moon. This vague standard was more metaphor than precise measurement, but no better system for predicting the effect of a lighting system existed. No one doubted that the new technology could provide considerably more light, but even the electricians themselves had everything to learn about how to properly distribute it. And so for years the American nightscape turned into an extended experiment in the quality of electric light, a festival of trial and error that provoked a sprawling public debate over what it meant to see in the dark, and how much that was worth.[20]

In the fading era of gaslight, each lamppost did little more than provide a beacon, a pool of light to draw the eye, guiding pedestrians from one lamp to the next along otherwise murky streets. Since gas mains rarely served working-class and outlying districts, pedestrians on these streets relied on much feebler oil lamps. Only parks and major boulevards featured the extravagance of multiple gas burners. The arc light promised so much more—while each gas lamp cast a light equivalent to sixteen candles, an arc produced a couple thousand. Of course, no one had ever experienced the collective effect of that many candles, a number so large that it lost meaning, so people searched for more tangible ways to describe the modern experience of abundant light. Under electric light, they wrote time and again, one could read a newspaper—some even tried to specify the type size and font that could be read at different distances. Others measured light by estimating the distance at which the face of a friend became recognizable, while some declared a street well lit if one could read a watch in the middle of it.

These were clear but crude forms of measurement, an immediate but subjective way of capturing how much the new light empowered the eyes. In their rough way, they described the quantity of light, but they left open the question about how much light people needed to use the streets at night. It was a truly remarkable modern experience to read the fine print of a newspaper in the middle of the road in the middle of the night—but was it an experience worth paying for?

Adding to the urgency of this question, many doubted the electric companies' claims about costs—not just the extravagant promises to deliver abundant light for a fraction of the old gas bill, but the increasingly more sober projections that electricity would cost "about the same," or perhaps a bit more. Crowds loved to gather to watch the new lights, but once they received their tax bills would they be willing to pay more for better lighting? The new technology sparked a debate that would go on for years, as electricians were forced to rely on trial and error, and calculations about the ultimate cost of the new light varied wildly, equations skewed by the twin variables of optimism and opportunism.

In order to deliver on their promise to deliver much more light for less cost, arc-lighting companies first attempted to install towers. Raising their powerful lights high above the city, they planned to create artificial moons, and calculated that a small number of these towers would deliver much more candlepower than the hundreds or thousands of gas lamps that burned on posts down at street level. An early enthusiast for this approach promised that with a handful of towers he could saturate the atmosphere above Mount Holyoke, Massachusetts, with an "artificial sunlight" so powerful that no additional light would be needed, indoors or out. He acknowledged a few problems with this plan. Since people had different bedtimes, some would want the lights extinguished sooner than others. And since the light would flow into each house whether or not its residents had paid their fair share, the town might need to banish those who refused to chip in.[21]

The arc light tower in San Jose, California.

Those lighting entrepreneurs who had less vision but more sense made no such claims for the tower system, but did suggest that towers would reduce the cost of lighting by 80 percent over any system that placed the light closer to ground level. The small city of Aurora, Illinois, was one of the first to try the experiment. Brush installed his powerful lights on a half dozen iron towers, each rising like a gigantic pencil over the city's rooftops. From this great height the Brush lamps provided intense light nearby and bathed the surrounding fields and "lonely outskirts" of the city with something like "full summer moonlight." In contrast to the arcs blazing high overhead, the gas lamps now looked more decorative than useful.

Declaring the Brush towers a "most brilliant success," one Chicago observer found Aurora's citizens "in a state of delighted enthusiasm over the splendid practical results."[22]

The tower system had its critics from the start. Some complained about the disorienting shadows cast by the brilliant arc lights. When trees or buildings blocked the light, the effect of moving from light to dark unsettled, a chronic problem in cities that had many hills. Thus the towers worked best in flat midwestern towns like Aurora that enjoyed long, wide streets set on a grid pattern. Some of these towns tried to further economize by erecting only one tower, but soon found that this left large patches of pitch darkness. Inevitably, they added more towers, replacing a single false moon with a constellation of brilliant stars.

The greatest test of the tower system came in Detroit. Eager to proclaim their city the best lighted in the world, aldermen contracted with the Brush Company to erect no fewer than seventy massive light towers around the city, each at least 150 feet tall. Brush offered to put the lights up at no cost, and promised to charge the city the same rate it was already paying for gas. But many Detroiters were skeptical. A former mayor with strong ties to the gas company argued that the tower system might work fine for a "prairie dog town," but would fail in a city with many trees and a more complex topography. Ignoring this warning, a majority of the city's leaders agreed to the experiment.

For months in the hot summer of 1882, Detroit residents turned out in droves to watch the massive towers go up. Even before Brush fired up his first lights, the towers proved controversial—some welcomed the thin iron spires as proof of the city's progressive spirit, but others found them an eyesore, especially those unfortunate enough to live near one, since each tower was braced by a wide and ugly spread of guywires and posts. Police arrested one man who tried to assert his property rights by attempting to chop down the wires near his home. Even without this vandalism some of the towers toppled, destroying the five-hundred-pound lights, splitting roofs, and scattering pedestrians. Growing in tandem with the arc-light business, the tower companies were just learning their trade.

When the Brush Company brought the lights online in August, this did nothing to quiet the controversy. The company had promised to deliver "a light equal to first-class moonlight," and it did so in many places. The effect was "picturesque and romantic," one observed. "The foliage is weird and beautiful. All places within scope of the light are bathed in the faint but fairy-like illumination of the moon in its first-quarter." In some spots it performed all too well, keeping geese and chickens awake all night until they began to die of exhaustion.

But the towers left other sections of the city, tucked behind hills and under shade trees, "absolutely without light." Foggy evenings thrust the whole town into darkness, and Detroiters could only speculate about the lovely sight that their lights must be creating as they shone down on the blanket of mist and soot that smothered the city. Even when light penetrated to the street level, many found themselves groping along sidewalks in an eerie gloom. "If people were accustomed to walk in the air," some complained, "the tower system would be very convenient, as the atmosphere is well charged." But down below, the deep shadows and sharp blue light left pedestrians "dazed and puzzled."[23]

Some visitors to the city found the whole effect poetic, like a fireworks display every night of the year, and they praised the city fathers for lavishly expending tax dollars to create such a sublime effect. But civic leaders took no pleasure in the compliment; they never intended to spend $110,000 per year "to inspire poetry." And they were embarrassed to find that other visitors, far from being impressed by the city's progressive spirit, found the whole arrangement to be crude. "It appears to me," one man observed, "that you are taking a very expensive way of getting a minimum of benefit from the electric lights."[24]

The Brush Company responded by weakly promising that all would look better in the winter months, once the leaves had fallen from the trees. And they frantically erected yet more towers, hoping to fill in the many gaps of deep shadow. But within a few years the city fathers conceded their error, while the *Detroit Free Press* declared the tower experiment "a flat failure." Detroit had not lost faith in the electric light, the paper insisted. "But the electric light that is believed in is one that will light the streets, not a few spots here and there, a back yard or two, and the firmament above." The towers came down—some dismantled, others falling in high winds.[25]

Some southern towns faced a distinctive tower problem of their own—a rash of runaway mules that, in their "fury," accidentally knocked down towers. In one case the mayor of Hannibal, Missouri, blew his nose "with so much vigor" that he spooked a mule, which then "dashed down the street, and ran against an electric light tower a hundred feet high, which at once toppled over." Musing over a series of these incidents, a New York editor could not decide whether this proved that southern mules were stronger than their northern counterparts, or that southern towers were weaker.[26]

Over the course of a decade, towers came down in most places. When San Jose erected its tower, boosters hoped the world would nickname it the "tower of light city." Instead they found that the massive structure was "practically useless" for lighting the downtown, and it survived for decades only as a curiosity. Boosters in Los Angeles had once boasted that their tower system made their city the best lighted in the country. Visitors, however, brought news from the East. "What you want," they explained, "is to have the lights down low, and to have them scattered at regular intervals." In a patronizing tone that must have stung the city boosters, these outsiders explained that it made little sense to light rooftops and the cosmos itself, while leaving sidewalks half lit. Within a few years, the city fathers began complaining that the electric light was "not what it used to be." A quantity of light that once seemed dazzling now looked dim and mottled.[27]

As one town after another abandoned the towers and brought their lights down to earth, they were also forced to lower their hopes that electricity would soon light their streets for a fraction of the cost of gas. Some city fathers were perturbed to find that in spite of many utopian predictions, electric light was often more expensive than gas, in some places by a good margin. But electricians had correctly surmised that once people experienced more and cleaner light, most would be willing to pay the premium. In a few towns, civic leaders abandoned electricity and turned back to gas, only to face the ire of voters who insisted on keeping the new lights at any price. The towers came down while demand for the new lighting systems continued to rise more quickly than the new companies could provide them.[28]

In addition to being more expensive, the system of placing more streetlights closer to the ground posed its own problems, including a growing public resentment against the poles. Many people hated the loss of light and air and their ugly intrusion on the landscape. Making matters worse, the hastily installed poles varied in size, often went askew, and soon bristled with advertising posters. And if not protected by bands of metal wire, they proved irresistible to horses and mules, which liked to nibble them into splinters. Landowners objected when crews showed up to erect the poles in their front yards, often hacking up shade trees to clear space for the lines. The courts spent years determining property rights in these pole disputes, most granting the electric companies the right of eminent domain since their service presumably served the public good. A few Luddites became local legends by fighting back, chopping down poles in the night. One New Jersey man went to jail for trying to cut down a pole when a lineman was still on it. In Baltimore, an "old man and a servant girl" tried the same thing, cheered on by their neighbors. He chopped till exhausted, then turned the ax over to her while he fended off the linemen. A woman chased a series of linemen away by threatening them with scalding water. Later she heard that local regulations barred the company from interfering with trees, so she planted a young plum in the company's half-dug hole. A Catholic congregation in upstate New York came to blows with a crew trying to put poles in front of their church. The sheriff only broke up the "row" by calling in the local militia. And in New York, the jeweler Charles Tiffany had better luck through the courts, winning a rare judge's injunction to prevent an electric company from erecting a pole in front of his store on Union Square. The judge agreed that the poles and wires discouraged customers and were "detrimental to the enjoyment by Tiffany & Co. of air and light." Despite Tiffany's victory, poles soon became a fixture on just about every town and city street in the country. Critics could only watch and complain as their streets became "impeded, lined, loaded-down and disgraced."[29]

In spite of these growing pains, a decade after Brush first unveiled his light on the streets of Cleveland, very few towns with a few thousand residents or more had no electric lighting to call their own. Quite often these systems were modest affairs, a dynamo or two firing a string of a couple dozen arc lamps—enough to light the main thoroughfare, maybe a park, and some factories and commercial buildings. Each arrived as a great curiosity and remained a source of civic pride. A booster might concede that his town was "a little out of the brisk course of events," but could still note proudly that "we have something to show in the electric line." In towns and villages across the country, local businessmen agreed that a lighting system would soon pay for itself, raising property values and attracting capital investment from those looking to bet their money on a "wide awake" town.

As a result, by the middle of the 1880s most town dwellers in America lived with the new light on a daily basis. But for many decades a majority enjoyed only a visiting relationship, without owning it themselves. They experienced the new light on the town's main boulevard or park, in a department store, theater, or hotel lobby, and perhaps in the office or factory where they worked. But at the end of the evening most returned to houses still lit by gas, kerosene, or oil lamps. Gas was nowhere wholly driven from the field in the first decades of this transition, especially since it became both cleaner and less expensive over these years. But electricians predicted that as Americans grew accustomed to the pleasures of this new product, demand would grow. Once a person's eyes were "trained up" to a higher level of light, as they put it, no one would ever feel quite satisfied living in a dimmer world.[30]

Three

Creative Destruction: Edison and the Gas Companies

In the months following his 1881 triumph in Paris, Edison worked with his team to introduce improvements and efficiencies in every aspect of his invention, as he prepared to install his first central power station in downtown New York. He overcame the political challenge of winning permission to dig up the city streets and the technical challenge of running eighteen miles of copper mains and wires, along with the fuses, meters, switches, and fixtures to serve more than a thousand customers. All of these elements were connected to his massive dynamos, which were housed in a modest four-story building on a run-down block of Pearl Street, centrally located to reach downtown Manhattan customers for half a mile in every direction. The overburdened station building's floors required special reinforcement before they could handle four large coal-powered steam boilers, six 240-horsepower steam engines, and the half dozen thirty-ton dynamos.[1]

In the late afternoon of September 4, 1882, the fire inspector signed off, and Edison fired up thousands of lamps in a square mile of lower Manhattan. After years of painstaking preparation and a half million dollars of invested capital, the system turned on without a hitch. In place of the usual dim flicker of gas, the bamboo filaments of the new lamps provided a "steady glare, bright and mellow, which illuminated interiors and shone through windows fixed and unwavering."[2]

Always a savvy publicist, Edison targeted those blocks of lower Manhattan that were home to many of the country's most important financial institutions and newspapers. In offices big and small, including the headquarters of two major newspapers, thousands of Manhattan's white-collar workers toiled for the first time under the incandescent light. Edison's team fashioned a variety of fixtures to hold these first bulbs. His desk lamps were based on the popular Argand oil lamps of the day; other bulbs attached to the adjustable wall fixtures of the now obsolete gas lamps; and some worked under simple ceiling lights—just a hanging wire, a reflecting shade, and the marvelous new bulb.

That bulb would soon become so ubiquitous, so mundane that it would become invisible. Those using the first ones marveled that they were "simplicity itself"—a glass globe shaped like a dropping tear, enclosing a slender horseshoe of glowing carbon. Glancing through downtown office windows, others saw droplets of fire. Just as Edison had predicted, his customers loved his bulb at first sight, mostly for all the ways it was not like gas. "No nauseous smell," one worker reported, and "no flicker." Clerks working under the light that night also expressed gratitude for the "decrease in the heat."[3]

Another surprising advantage—"you turn the thumbscrew and the light is there . . . no matches are needed, no patent appliances." Although Edison's incandescent lighting system was one of the most sophisticated pieces of technology yet created, all the complexity had been engineered out of sight, invisible to the consumer. Each light socket, as one of his new customers explained, "contains a key whereby the lamp may be turned on or off at pleasure." Oil lamps and candles required wick trimming and soot cleaning, while gas burners demanded even more technical skill from consumers, who had to adjust meters and burners in addition to regular cleaning. But the electric light required no maintenance, while the source of power hummed out of sight, sometimes many city blocks away. As far as the customer was concerned, the bulb worked for about six hundred hours, until it either broke or began to blacken and dim. Then an electric company worker could replace the expired bulb in a minute or two. Thus Edison's system realized from the start an essential feature of any modern invention aiming to win a mass market: it was safe enough for a child, and simple enough for all to use; not just foolproof, as Edison said, but "damned fool proof." Although a functioning light bulb represented the culmination of decades of scientific insight, inventive genius, and technical skill, consumers encountered it as a mass-produced object—not quite cheap at fifty cents or a dollar, but an expendable item to be tossed in the trash when it broke.[4]

Edison's public triumph was tempered only slightly by two embarrassing glitches, each involving private systems for his wealthy patrons. Eager to be among the first to enjoy incandescent light, William Vanderbilt arranged to have Edison's lights installed in his Fifth Avenue mansion. On the first evening they worked fine, until a wire interacted badly with metallic thread woven into the wallpaper and began to smolder. Horrified to discover that her husband had installed one of Edison's mysterious dynamos in her basement, Mrs. Vanderbilt ordered the whole thing removed, a story gleefully retailed by the gas companies. A bit later J. P. Morgan installed his own private system, including a light specially designed for his desk—the world's first desk lamp. When it was turned on for the first time, however, the lamp's faulty connections caused a fire that left the desk a charred ruin. But Morgan remained an enthusiast, keeping his Edison lights and ignoring his neighbors' complaints about the dynamo that rumbled in his yard.[5]

In spite of those setbacks, on that September evening Edison stood in his engine room on Pearl Street, sleeves rolled up and in what one reporter called "a high state of glee." Only weeks before, leading electrical experts in Europe claimed that Edison would fail. As he enjoyed his vindication, the gas company serving downtown New York was deluged with requests from longtime customers demanding the immediate removal of their burners and their hated gas meters.

The arc-light companies had already made great inroads against the gas industry and seemed bound to win the battle for the valuable contracts to light the streets. While gas companies fought this development with all their formidable power, they took comfort in the fact that the intense arc light could never compete in the far more lucrative market for interior lighting. Now Edison threatened to take this business too. The demand grew so briskly that one of Edison's supporters predicted that the infant incandescent lamp would soon "come to manhood and strangle gas with one hand and petroleum with the other."[6]

Both Edison and his rival electricians sold stand-alone systems, single dynamos that fired a string of lamps, enough for a large house, store, or ship. But after the successful test of his Pearl Street station, Edison hoped to move forward with his much grander vision for an electrical grid, installing his central system in the urban core of every major city. Each territory offered a potential market of tens of thousands of lamps for office buildings, theaters, and the private residences of the elite. Edison's company planned to sell its equipment to a local utility, which would pay royalties and assume

Turning coal into light at Edison's first central station on New York's Pearl Street, 1882.

responsibility for finding and serving its customers. Once free from the obligation to oversee the daily operation of his New York power station, Edison devoted his time and resources to improving every aspect of his system. In order to vertically integrate the many technical components of his Pearl Street station, he founded a series of interlocking companies. Now he supervised their work in developing and manufacturing dynamos, underground conduits, fixtures, and bulbs. Edison set out to apply this strategy not only to conquer the United States, but to sell his invention around the world. He arranged similar partnerships with local utility operators in major cities in Europe, Asia, Central and South America, and Australia.[7]

. . .

In their attempt to win customers over to electricity, the new electrical entrepreneurs enjoyed an advantage—many people shared their eagerness to throttle the gas companies. Gas men had been "humbugging the people long enough," many complained, abusing their monopoly to wring hefty profits. Plenty of irate customers had a story to tell about their "thralldom to the gas man" and the companies' indifferent service. Much of this ire focused on the gas meter, a mysterious device that determined each customer's bill. Judging from cranky letters to the editor in many nineteenth-century papers, no gas customer ever believed that this "demon meter" offered an honest accounting. Critics described the meter as a mysterious box that the company installed in each house for the purpose of printing money. Not long before, the gas companies had been welcomed as the vanguard of progress, their business nurtured by lucrative city franchises. Now they had become the villains, depicted in many forums as greedy curmudgeons and connivers who put their own interests ahead of progress and the common good.[8]

Those left holding shaky gas stocks looked to Andrew Hickenlooper as the great defender of their economic interests. As head of the Cincinnati Gas Company and sometime president of the American Gas Association, Hickenlooper became the most vocal and visible critic of the new technology. His long and successful business career began at the Battle of Shiloh in 1862. A military engineer in Ulysses S. Grant's army, he acquitted himself well enough in that narrow Union victory to emerge as something of a local war hero. Returning to Cincinnati after the war, he found a city that was growing by leaps and bounds, and thanks to his army connections he landed a post as a city inspector of new building projects. The trustees of the Cincinnati Gas Company felt it would be more con-

An 1878 lithograph praised Edison for "Light thrown on a dark subject (which is bad for the gas companies)."

venient to own Hickenlooper than to be regulated by him, and soon offered him a nice job as vice president.

At the time Hickenlooper knew nothing about gas, and his new partners clearly wanted him more for his political connections than for his engineering expertise. But the young man threw himself into the task with an intensity that amazed his friends and annoyed his foes for decades to come. He mastered the technical challenges of producing and distributing gas and developed many improvements in the system. He installed new docks on the Ohio River and furnaces around the city. As Cincinnati expanded into the era's new streetcar suburbs, so did the gas lines. Hickenlooper even cleaned up incompetence and corruption in his own company, saving it from the threat of a city takeover. By 1872, he had emerged as the autocrat of a thriving utility, powerful enough to beat back challenges from a number of rival gasworks. Whenever the city considered granting contracts to any of Hickenlooper's challengers, he warned that the ensuing competition would only produce bad service and higher prices. He famously rammed that point home every time the company's interests seemed threatened, overwhelming his foes with a barrage of statistics.

Hickenlooper poured his lifeblood into the Cincinnati Gas Company. In an era defined by conflicts between capital and labor, he learned to hire sober and complacent German immigrants. During the great labor violence of 1877, rumors spread that strikers in Cincinnati planned to take over his works, plunging the city into darkness and chaos. Moving through the mob in disguise, Hickenlooper made it to his plant in time to hand out revolvers to his loyal workers, repulsing a threat from a group he called "the toughest lot of human beings of both sexes I have ever encountered." With similar verve, he parried attacks from city officials and newspapermen who, in his view, courted public support by "indulging in all sorts of unwarranted abuse of the gas interests." One author, for example, lampooned General Hickenlooper as the blustering "Commodore" of the city's "4-C's—The Celestial Coal Smoke and Coke Company," a man who treated the city as his personal fiefdom. In reply, Hickenlooper could only shake his head and say, "Men otherwise reputable appear to lose all sense of truth and justice when they become engaged in any controversy with the gas companies."

Hickenlooper's political power grew along with his fortune. He never missed a reunion of Ohio's Union veterans, and parlayed his pull with his fellow soldiers into a stint as lieutenant governor. Whenever a large parade marched down the streets of Cincinnati, chances were good that Hickenlooper rode along as grand marshal.

Clearly Hickenlooper was a force to be reckoned with, but was he powerful enough to stop the march of technological progress in its tracks? He felt sure he could, and he led the charge against what he liked to call "the electric light scare." When the Brush Company first demonstrated the arc light in a downtown Cincinnati square, the event spread panic among local investors, forcing Hickenlooper to use his own fortune to shore up the price of his company's stock. Months later, when Edison made his premature announcement that he had solved the "puzzler" of incandescent lighting, the intrepid Hickenlooper once again donned a disguise, traveling to New York to impersonate a stock speculator so that he could investigate the matter firsthand. Returning to Cincinnati, he assured investors that the quest for a viable incandescent lamp was an old story and that Edison offered nothing new. At the time, this evaluation was as true as it was self-serving. He hedged his bet, however, by working to expand the market for gas in fields aside from lighting—a strategy that would in the end save the gas companies from utter destruction. He hired Miss Dodds, recent graduate from a Boston culinary school, to give public demonstrations on the advantages of cooking with "gaseous fuels."

Following Hickenlooper's lead, gas companies around the country fought back against electricity by consolidating their operations, lowering prices, and introducing new, more efficient technologies. They also called in favors from their political allies, ran special deals during "gas lighting week" each spring, and kept up a steady drumbeat of propaganda about the dangers of electricity. "Wide-awake" gas men assured each other that the recent boom for electricity was just a passing fad and that they were bound in the end to win the struggle between "the gas meter and the dynamo." Through all this, Hickenlooper continued to insist that if the city fathers granted contracts to rival lighting companies, both gas and electric, this would introduce chaos in the marketplace, ruinous competition that would ultimately produce higher prices and worse service. In the parlance of the economic debates of his day, Hickenlooper insisted that public lighting was a "natural monopoly," and that it should be his.[9]

A more immediate argument against electric lights came not from chaos in the markets but chaos in the streets, where rival arc-light companies began to weave a thick web of poorly insulated high-tension wires overhead. Eager for more light, and sometimes private kickbacks, city politicians granted franchises to the various electric companies with no attempt to impose order or even minimal safety standards. And in a number of cities, hastily organized companies wasted no time asking anyone's permission. They just tacked their wires on whatever poles and building facings served their purpose, sometimes taking a cost-saving shortcut directly over some poor homeowner's roof.

The electric companies strung these high-tension wires along streets already thick with wires for telephone, telegraph, fire and police alarms, and stock tickers. In dense urban intersections a pole might carry as many as two hundred different wires. Those lines were unsightly, but used a moderate current that posed no danger. All this changed when electric companies added their powerful and poorly insulated high-pressure arc wires to the mix. Hastily tacked up, these often broke loose and fell across the web of other wires overhead. Traffic stopped and crowds gathered as wires burned and sparked, sometimes flailing like "fiery serpents." In the 1880s, city residents often enjoyed these "free fireworks displays," but sometimes the results proved more serious. Once in contact with broken or sagging arc wires, harmless telegraph, fire alarm, and telephone wires delivered awful, even deadly shocks. At other times they burned and melted, causing numerous fires. The firemen who came to the rescue faced not only the risk of the blaze but also the danger of electrocution. During a serious fire in St. Louis, for example, firemen found themselves unable to retreat from a wall about to collapse, since they felt hemmed in by the deadly wires. The crackle of exploding wires spooked their horses, and at one point their ladder truck became so "completely charged from fallen wires" that none dared approach it. Electric light companies delivered on their promise to avoid the dangers of burning gas, but only by introducing some nasty hazards of their own.[10]

For late-nineteenth-century city dwellers, the sky overhead became increasingly ominous, thick with wires that might pour down a man-made lightning bolt without warning. "The overhead system," one medical journal declared, "is a standing menace to health and life." Every week the papers ran stories of this very modern form of sudden death. A Memphis man tied his mule to an iron lamppost that had been accidentally electrified; the powerful current knocked the screaming mule off its feet, and when its owner came to the rescue he leaned against the post himself and was instantly killed. An Italian fruit vendor in Greenwich Village slipped while cleaning the roof of his shop, touching one of the dozens of electric wires converging there. He probably died instantly, but a horrified crowd gathered to watch his corpse sizzle, as long flames shot from a wire lying across his neck. Fearing their own electrocution, the police dared not go near until an electrician arrived to cut down the wires. A pole painter in Massachusetts slipped, grabbed the nearest wire with both hands, and found himself in the throes of eight hundred volts; only the heroic action of his partner saved his life, though his hands were horribly burned. Children playing in the streets fell victim quite often, as they enjoyed the sport of reaching for dangling wires or climbing on street poles. The old wires just tingled, but the new ones killed.[11]

Wires that had been hastily thrown up came down even more

The introduction of electric lighting turned the "wire nuisance" into a deadly threat, 1881.

quickly in bad weather. Electric wires laden with icicles put on a fascinating but terrifying light show, making electrical connections between wires that erupted in showers of blue sparks. The sight drew crowds into the dangerous streets, who then dashed for safety each time one of the "demoralized wires" fell to the ground. During one storm in St. Louis, wires scattered across the city streets killed two horses, taking their heads nearly off, knocked many pedestrians senseless, and turned one poor woman's pet dog into "a corpse in a jiffy."[12]

Most of the electric light's victims worked for the companies. At a time when the properties of powerful currents were barely understood and safety standards for the industry were just being invented, these sensational accidents provided lurid fodder for the daily papers. Some workers fell from light poles and crashing towers; others suffered when repairing wires that they assumed were not live. Arc lights required daily maintenance, as the carbon wicks burned out each night and the globes needed cleaning, a job that brought these lamp trimmers perilously close to live wires. In a particularly gruesome incident, a New York man hanging telephone wire in the pouring rain accidentally touched a poorly insulated arc wire. Thrown from the pole, he became entangled in the wires, hanging head down in midair, "perfectly helpless and insensible." He was rescued but suffered awful burns, including a toe charred to the bone.

Many more were struck down while working around the dynamos, accidentally completing a circuit that sent the powerful current through their bodies. Papers described these men as suddenly paralyzed, "powerless to unclench" their hold on the wires or the dynamo. Most suffered in excruciating silence, issuing at best a "faint and unnatural cry." As their flesh burned, coworkers sometimes fled the room in panic, expecting an explosion; the more fortunate were rescued by colleagues who braved a terrible shock themselves in order to break the circuit. At a time when prison reformers were exploring the use of electricity to execute prisoners, one editor suggested that death row inmates should simply be apprenticed to work for an electric light company—sooner or later, the job would carry them off.[13]

Every week the papers carried a new story about "the latest victim of the electric current," along with many more about close calls. When an elevated train in New York collided with a low-hanging arc wire, passengers heard powerful crackling sounds and saw showers of sparks falling from the roof. They got away with nothing more than a strong tingle. Only a week later, the driver of a horse-drawn trolley was not so lucky when his rig ran into a powerful arc wire dangling just five feet above the ground. Doubled over by a powerful jolt of current, he was thrown across the street. The crowd that soon gathered assumed the man was dead, until he "contradicted this by howling." Luckier than many, he suffered only a bad burn on his knee and a bout of what the newspaper called "nervousness." Electricity was proving to be as treacherous as it was helpful, and *Scientific American* declared the plague of "vagrant electricity" to be "one of the most serious and alarming of city evils."[14]

Advocates for the electrical industry had to acknowledge the dangers but insisted that most papers greatly exaggerated the risk. Some did so, they charged, because electrical death seemed more novel, and thus more newsworthy, than the city's usual allotment of crushings, drownings, and knifings. But electricians also suspected that some editors fueled the public's fear of electricity at the bidding of their financial backers in the gas companies. Hickenlooper was known to keep his own file of stories about electrical disasters, and enjoyed sharing it with any reporter who could be paid to listen.

The electrical trade journals fired back, never missing a chance to report on some of the many victims of accidental (or intentional) gas poisoning. "If electricity has slain its tens," one editor wrote, "gas has certainly suffocated its tens of thousands." Drunks quite often blew out the gas before going to bed instead of properly turning the stopcock, a fatal mistake also made by many rural people visiting the city for the first time. When underground mains broke, gas sometimes leaked into cellars, poisoning those who slept there. Many suicides chose asphyxiation by gas. After an explosion at a St. Louis gas company killed two men, the *Electrical World* lectured that "more damage was done by that one explosion than has been caused by electric lighting for the whole of the last year." The friends of electricity correctly noted that in spite of widespread public anxiety, electric wires caused only a small fraction of accidental deaths. And they blamed most of these accidents not on the technology itself but on worker negligence or the reckless stupidity of the young. Desperate to defend their new enterprise, they dismissed these searing public executions of the innocent as "the penalty of folly."[15]

Electric light journals also gathered and circulated the wider universe of complaints about gas lighting. In confined spaces, gas often heated rooms to intolerable levels, consuming oxygen and leaving noxious fumes in its place—some compared it to introducing an open sewer pipe in every room. Burning gas spewed steam, sulfur, and ammonia into the air, which acted as a corrosive on fabric, leather bindings, and paintings and covered every object with a filthy coat of oil that served as a magnet for dust. Valuable books crumbled on library shelves, ceilings were blackened with soot, and houseplants withered and died. In a brief time, the by-products of gaslight took a terrible toll on the walls and tapestries of Westminster Abbey, the tombs of Egypt, and priceless works of art in museums around the world. Those rich enough to afford it found themselves redecorating their rooms each year in order to keep up with the destructiveness of their lamps. Once, many city dwellers had considered this stench, rot, and stifling air an unavoidable by-product of progress, worth suffering in order to satisfy the craving for more light. But now, as one electro-journalist boasted, the public "has almost as much contempt for gas as they had for oil lamps. Old friends fade before brilliant newcomers."[16]

Within a few years, even Andrew Hickenlooper, the gas man's gas man, saw the handwriting on the wall and began to suggest that his company was not in the gas business but in the *lighting* business. As such, he lobbied state and local governments for the legal right to expand, adding electric light to the company's offerings. He even conceded that electric lighting had been good for the gas business; as people grew accustomed to the luxury of strong light, they burned more gas if that was all that was available. Of course, his longtime rivals in the electrical business crowed. "Time was when General Hickenlooper could not say anything too strong against the electric light," the *Electrical World* taunted. "But Saul is once again numbered with the prophets, and we are able to present . . . a signal proof of the thoroughness of . . . Hickenlooper's conversion." A nimble warrior, the general decided the electric light was not so bad, especially if he could win the exclusive right in Cincinnati to provide it.[17]

Hickenlooper's concession was only a strategic retreat, and the struggle between gas men and electricians would continue for decades. Electric lighting technology remained imperfect and more expensive, while gas technology rapidly improved. But many of the fledgling electric companies now faced "new conditions." As many of the well-capitalized and politically influential gas companies followed Hickenlooper's lead and branched into the electric light business, this provided a moral victory for the new technology but formidable competition for many of the smaller electric pioneers.

A leading British electrician who toured the United States in the early 1890s expressed surprise at Americans' indifference to the new technology's dangers. "The accidents in the States are terrible in their number and fatality," he reported back home. Another wondered how Americans could endure it. "Electric light wires everywhere cross your path," she noted, "making accidents extremely liable." That, along with new thirteen-story skyscrapers, evidently built on thin air, made the whole city seem as temporary and liable to burst as a soap bubble.[18]

Yet even as British electricians insisted that they "do these things better," some conceded that they had "everything" to learn from their American counterparts, whose reckless enthusiasm for electric light had produced remarkable results in a short time. "A visit to the United States is something like charging an accumulator," one concluded. "It stores the visitor with energy."

Through the 1880s, Europeans could only wonder that so many of America's small towns, big stores, rail depots, and sidewalks were already brilliantly lighted. Wealthy Europeans installed private lighting systems in their own homes, but American entrepreneurs sold thousands more of these private systems, while towns and cities erected central stations, capable of delivering light to a much wider range of customers. A decade after Brush's early experiments, and only five years after Edison lit up lower Manhattan, American electricians noted proudly that "while many of the largest cities in Europe are still without central stations, small villages in all parts of this country are putting in plants and operating them profitably." A British expert touring American cities in 1884 noted with some amazement that New York's main thoroughfares were "entirely lighted by electricity." Returning to London, he traveled a similar stretch of that city "without seeing a single electric light." After seeing "so much brilliance" in American cities, he confessed that living under London's gas lamps was a bit depressing.[19]

Struggling to explain America's rapid embrace of the new technology, European visitors assumed it had something to do with the raw, improvisatory nature of American cities, where considerations of safety, aesthetics, and tradition were no match for the seller's pursuit of profit and the buyer's fascination with all things shiny and new. Americans had scarred their streets with a mess of wires and poles, thrown up so hastily that they rained deadly current down on hapless pedestrians and routinely killed and injured electrical workers. In spite of these hazards, Americans had deemed the electric light to be a "good thing, and if once a thing is proved to be good in America, it is immediately taken up. 'Progress' in that country is the word."[20]

The Brooklyn Bridge at night, 1890.

Four

Work Light

Not long after New York erected its great Brooklyn Bridge, the social reformer Helen Campbell visited a family that lived in a squalid tenement apartment, tucked hard against one of the bridge's massive stone piers. Except for a brief glimpse of the sun at noon, their rooms had been cast in permanent shadow, and even that small amount of light disappeared when neighbors hung out their washing. To see in the daytime they needed to burn their oil lamp, an expense they could not afford. The Brooklyn Bridge, one of the great structures of the modern age, had turned them into cave dwellers.

All of that changed each night when the city lit the bridge's powerful arc lamps, casting a blaze of artificial light into the family's small rooms. And so they adapted. Each sunset became their dawn, when the bridge's light roused them out of bed. They worked all night by this "hard" light, sewing trousers for a sweatshop, and when the bridge's lights went out at sunrise they retired to sleep through the false night of its shadow. Surveying the topsy-turvy nature of their lives, Campbell believed, "Never was a deeper satire upon the civilization of which we boast. Natural law, natural living, abolished once for all." A heartless industrial order had forced this "family under the bridge," as she called them, to forget the very meaning of sunshine and accept the modern world's new substitute, an artificial light that "blinds but holds no cheer." The family saw things a bit differently. "We would like to see the sun once in awhile," one of them explained. "But we go out for that, and it's better than nothing." Whether this family's upended schedule was a minor inconvenience or a mockery of American civilization, their story suggests the disruptive effects of electric light on people's lives, its power to scramble the timeless verities of day and night, forcing many to adapt as best they could to modern life's new, man-made chronological order.[1]

For all of human history the rhythm of night and day exerted a powerful influence on how people arranged their lives. Nature seemed to intend the daylight hours for toil and the night for rest. This rough rule of thumb made sense in early America's agricultural society but came under challenge during the nineteenth century's illumination revolution. When New England textile mills began to extend the workday in the 1840s by lighting up candles and oil lamps, the female operatives denounced this "outrageous custom," one that they found "not only oppressive but unscriptural." As some of them pointed out to their bosses, God himself had done no night work when he made the Creation.[2]

Gaslight, and the new pressures of industrial production, had already begun to blur the line between day and night, but now the more powerful electric light threatened to erase the distinction entirely. Some joked that in the near future authorities would need to fire a cannon to announce when "day" and "night" ended; otherwise, who could tell?[3]

Instead gas, and then electric, light complicated the primordial bifurcation of night and day by adding a third option, an illuminated evening that mixed elements of brilliance and shadow, looking and feeling like nothing any human had experienced before. These lit hours between sundown and bedtime became a new piece of time, chronological territory fought over by labor and capital, and colonized by a new breed of entertainment entrepreneurs.

In many industries, owners embraced electric light's economic potential, eager to keep their factories, mills, and shops open and their goods moving. Their workshops had required an enormous outlay of capital for expensive machinery and expanded facilities, including the new electric motors that were revolutionizing the production process in many fields. To earn the best return on that investment, owners needed to keep those machines running as much as possible. As Henry Ford put it some years later, "expensive tools cannot remain idle. They ought to work 24 hours a day." Many capitalists realized quite early that with a modest investment in an electric lighting system, they could effectively double a factory or mill's productive capacity—but only if workers would cooperate by working long into the night.[4]

Anticipating this trend, some friends of the working class predicted that electric light, which seemed to promise middle-class consumers nothing but pleasure and convenience, would only bring further misery to the working class. Forced by poverty to work whenever bosses offered them the chance, and even to send their children into the factories, these industrial workers found some small measure of protection in the darkness, the one time when the "taskmaster" could not demand their toil. Now a flood of inexpensive artificial light into fields and factories threatened to erase this natural, God-given safeguard against exploitation. Facing the prospect that electricity would make possible the perpetual workday, one British commentator wondered aloud "whether continuous light will altogether add to the happiness of man."[5]

. . .

In the early decades of the Industrial Revolution, workers relied on sunlight as their primary light source, often setting their benches as close as possible to the factory's tall windows. Some employers tried using candles and oil lamps to extend the workday, especially during dark winter months, but they avoided night work whenever possible; efficiency declined in dim artificial light, especially when workers had to give up one hand to hold their own lanterns. The introduction of gas offered a stronger light for night work and its central supply system made possible the illumination of entire workshops and factory floors, supporting a production process that was increasingly complex and integrated. But gaslight remained expensive, still caused fires and explosions, and—since it was not portable—proved useless in many work situations.

Where electric lights were installed, workers immediately noted its advantages. The new light relieved them from the nasty smell and oppressive atmosphere of burning gas or oil lamps. In addition to having clearer heads, they enjoyed clearer vision, no longer deceived by a flame's yellow flicker. In the post office, for example, some mail sorters found that they could cast aside their spectacles, no longer forced to squint in gaslight's "semi-darkness."[6]

In some industries, powerful, precise, and tireless machines now handled work that traditionally had been done by skilled artisans. Mass production arrived at different times and changed each industry differently, but tended to "de-skill" craftsmen, replacing them with unskilled and lower-paid factory operatives. However, mechanization also created a demand for new skills, the sharp eyes and sure hands required to maintain and run complex machines. These workers welcomed the electric light, which not only improved the quality of their work but also helped to prevent industrial accidents. At a time when powerful machines came without safety devices, the best way to prevent the loss of life and limb was to keep a keen eye on one's work.[7]

In some trades the more powerful light proved a useful tool of production, making work not only safer and faster but also more accurate. Thanks to electric light, for example, textile workers could now see the true colors of their handiwork. Mapmakers such as Rand, McNally & Co. found that it greatly improved their printers' ability to create consistent tints of yellow, green, and blue. In this way, this invention contributed to a revolution in the quality and quantity of graphic material in the Gilded Age, satisfying what one historian has called "the expanded public's craving for visual information." With the help of stronger light, along with the steam-powered press and new techniques of chromolithography, printers catered to a growing market for richly illustrated magazines, children's books and pulp novels, advertising cards and colorful packaging, a new world of visual stimulation that played a central role in the development of the era's mass consumer culture.[8]

Those newspapers not in the pockets of the gas companies became early adopters of the new technology, and often its greatest fans. Many newspaper editors thought of themselves as patrons of their city's future, and they used their considerable influence to encourage residents to subscribe to electric lighting contracts and to turn out for early exhibitions. Installing arc lamps, and then incandescents, newspapers made their offices into showrooms, inviting readers to stop by for a look. Newspaper tradesmen loved the new lights, which eliminated the stifling heat and oppressive atmosphere of gaslight in their cramped offices and provided much better working conditions for typesetters, who toiled over fine print late into the night, racing to arrange the next day's columns. One publisher felt sure that the new light precipitated a moral reform in the profession. In the gaslight era, long hours in close quarters left printers "hot and excited," inspiring a thirst for drink that made their profession notorious as a drunkard's trade. All that ended, one boss noted, when electric light was installed, producing a "great change in the habits of the printers." The editor of a Washington, D.C., daily captured the great improvement that electric light brought to office workers, celebrating his paper's "absolute emancipation from the slavery endured while an enforced patron of the gas company. . . . The smoky-yellow glare of the villainous fluid which the gas company kindly permits to ooze through its kerosene-tar clogged pipes only exists in memory, as the recollection of a nightmare."[9]

The transportation industry was another early adopter of the electric light, using it to extend the reach and value of the era's powerful new railroads and steamships, culminating by the early twentieth century in a twenty-four-hour-per-day distribution network—a foundation of the modern industrial economy.

Long before the arc light found any other commercial application, the British pioneered its use in lighthouses. The earliest relied on battery power and thus were quite expensive, but this was a price deemed worth paying to protect so much life and property. On both sides of the Atlantic, inventors worked to create electrified buoys that might replace the old oil lamps. New York authorities tried to minimize nautical accidents by lighting as much of the harbor as possible. In addition to the city's lighthouses, both the Brooklyn Bridge and the Statue of Liberty featured electric lights that served a dual purpose as navigational aids. But New York's greatest attempt to illuminate its waters was a massive arc-light tower at Hell Gate, a treacherous strait on the East River. The sight was more impressive than effective, however, as pilots complained that they were blinded by the glare, a problem also faced by mariners passing by the Brooklyn Bridge and other coastal arc lights. Baffled as much

Brush arc lights on the East River's treacherous Hell Gate.

by the light as they had been by the dark, some tugboat operators still refused to go near Hell Gate at night.

The Industrial Revolution that transformed the American economy in the Gilded Age set the world in motion: millions of immigrants crossed oceans to find work, while manufacturers and farmers scanned the globe in search of new markets for their goods. As the nation's foreign and domestic trades grew exponentially, so did the traffic in its harbors and shipping lanes. By the early

1880s, over twenty thousand ships from around the world sailed into New York harbor each year. The coast teemed with small ferries and massive transatlantic steamers, local barges and canal boats hauling coal and wheat, sloops from many nations carrying everything from bricks to ice to tropical fruit, fleets of fishing boats and pleasure craft, and many more vessels. Night brought particular hazards to these crowded waters, and the traditional safeguards of oil lamps and watchmen with whistles failed to prevent a growing number of disasters. Modern steel-hulled, steam-powered ships were bigger and faster than ever, but in dark, crowded seas they were also more dangerous to one another. Through the 1880s, newspapers regularly reported on harrowing collisions, complete with vivid accounts from those who survived, and long lists of those who did not. As one editor put it, "Almost weekly, hundreds on hundreds are being drowned like rats in a cage." He offered a modern solution to this modern horror. The "electric ray" should be required equipment on every ship, replacing the feeble "firefly" of oil lanterns with a light powerful enough to "pierce the thickest fog or tempest for many hundred yards."[10]

After a midnight collision in the English Channel drowned many of the passengers on an American steamship in 1878, *Scientific American* likewise called on inventors to create a viable electric light for commercial ships, a breakthrough that promised to be both lucrative and lifesaving. Not long after, American inventor Hiram Maxim answered that call, developing a nautical spotlight keen enough to "cut out a way for itself through the fog," one of the earliest of many attempts to use electric light to make night travel safer on the seas.[11]

Railroad operators and inventors made similar attempts to adapt electric light, as part of their effort to create a round-the-clock transportation network. Late-nineteenth-century railroads provided the crucial link that enabled the growth of powerful and efficient corporations, able to move natural resources in enormous quantities and to deliver goods across a national market. Passenger travel also soared, serving an increasingly mobile society. As one American professor exulted, "Our railroads carry us hither and thither on the earth with somewhat the facility of spirits." More than any other technology, the railroads broke down barriers of time and space and, as one historian puts it, allowed "movement at any time and in any season or type of weather."[12]

Railroad companies had strong financial incentives to keep their trains running at all hours, and were eager to resolve any technological barriers that stood in the way of faster and more efficient service; this inspired inventors to look for solutions on a number of fronts, such as George Westinghouse's invention of the air brake, which allowed safe travel at greater speeds. But the darkness posed a different problem for the rail companies, a hazard that limited the use of the rails at night and scared away passengers. "Many people are averse to traveling on railways by night," *Scientific American* noted in 1864, "having the impression that there is greater liability to accident." The journal pointed out that night travel was not more dangerous, but only because traffic was minimal and the trains moved cautiously. Fewer were killed, because fewer dared the trip. Railroads were eager to invest in a headlamp that would make night travel safer and faster.[13]

In the 1830s, an inventive South Carolinian tried to solve this problem by rigging a basket of flaming pine-knot torches to the front of his locomotive. Later inventors experimented with more reliable oil, kerosene, and gas lamps. Concentrated by parabolic reflectors, these lights could at best cast a beam one thousand feet down the track. Such a light might warn pedestrians to clear the tracks but had "practically no value for the engineer" since it took twice that distance for a train to stop. In the push to use rail lines at full capacity, more speed demanded more light.[14]

Not long after Brush unveiled a reliable arc light, other inventors raced to be the first to successfully adapt this technology to the train headlamp—all recognizing the economic value of an electric beam capable of surviving the jarring vibrations in a locomotive. Early models proved inconsistent, sometimes plunging fast-moving trains into the darkness, but when they worked they could illuminate the track for a half mile or more ahead, allowing safe travel at higher speeds. The technology continued to evolve rapidly in the late nineteenth century, including the development of twin beams, one to illuminate the track and another to send a warning signal that was visible up to twenty miles ahead. But many railroad men resisted the electric headlamp. Some traditionalists feared that bringing a dynamo on board increased the danger of being hit by lightning. More commonly, though, the engineers complained that the powerful headlamps of oncoming trains blinded them. Even worse, the lights made track signals difficult to read, undermining the value of an electrical device that remained the most effective safeguard against accidents. Attempts to replace the arc headlamps with less powerful incandescent bulbs failed, since the filaments proved too fragile to endure the vibrations. The technical challenges continued to vex inventors and railroad men long after states began, in the early twentieth century, to mandate the use of high-powered electric lamps on all locomotives—choosing the safety of those who crossed the tracks over the concerns of those who ran on them.[15]

The light arrived only slowly in passenger cars as well. Before the Civil War, when night service on passenger trains remained rare, large candles were used, one at each end of a railcar—a dim light improved in later years by the use of oil and kerosene lamps. Some rail companies installed gas lines, though these brought with them all the usual complaints about noxious, overheated air. Still, rail companies resisted making the investment in electric light long after it became standard equipment on ships, and in most other public spaces, because the formidable technical challenges drove up the cost. Inventors provided a half dozen solutions, combinations of portable dynamos and massive storage batteries. One even developed a slot machine for railcars that provided "a soft and abundant light" for only a nickel, designed to "enable one to pass the time quickly and agreeably when traveling," though the gimmick never caught on. For many years incandescent light remained an expensive luxury, reserved for passengers on "specials" who were willing to pay a premium for a superior light. When President Grover Cleveland toured the country by rail in 1887, for example, his party traveled in three adjoining railcars elaborately equipped with a barber chair, pipe organ, and smoking room, all illuminated by an incandescent lighting system that took up a large part of one of those cars.[16]

Steamboats also adapted electric light for night travel, though some river pilots resisted the change, finding the light's dazzling effect too disorienting. One tourist traveling on a Louisiana steamboat in the early 1890s reported that the pilots still preferred to pick their way through the dark, using their powerful searchlights only when trying to find a landing. The effect of these "great shafts of daylight" thrown against the "blank wall of night" turned evening travel on the river into a real-life magic lantern show. "Each light cut a well-defined path through the night," as he put it, "and when it picked out a grove of trees or a clutter of negro cabins or a landing, it created a veritable stage picture." On other inland bodies of water this novel sight became a tourist attraction, with passengers flocking to "search-light excursions."[17]

An enthusiast for new inventions, Mark Twain considered the use of electric light on steamboats to be an enormous improvement over the "flickering, smoky, ineffectual, pitch-dripping torch baskets" used when he had been a pilot in the 1850s. As Twain recorded in *Huckleberry Finn*, before the electrical age night travelers on the

Mississippi faced the dangers of hidden snags and sandbars, and collisions with other boats and rafts. "All that is changed now," he noted in his account of a river journey from St. Louis to New Orleans in 1882. "You flash out your electric light, transform night into day in the twinkling of an eye, and your perils and anxieties are at an end." Twain considered the sight of the New Orleans levees at night, a five-mile crescent lined with electric lights, to be the highlight of his entire trip. For the shippers and stevedores of New Orleans, those lights allowed the port to operate efficiently every hour of the day. But to Twain, their value was more aesthetic than economic. "It was a wonderful sight," he thought, "and very beautiful."[18]

Electricity was changing not only the way goods were produced but also how they were sold. Over time, merchants became increasingly sophisticated masters of the light's power to seduce customers, but in the early years all it took to draw a crowd was the naked glare of the light itself, and the more of it the better. As in so many other aspects of the development of the Gilded Age department store, John Wanamaker led the way. When he opened his "Grand Depot" on Philadelphia's Market Street in 1877, he had used natural light from windows and skylights, supplemented by massive gas chandeliers. And he experimented with light to lure customers, inviting Christmas shoppers to enjoy a "grand illumination" produced by a lavish display of gas jets and colored reflectors. While the public raved at the sight, Wanamaker regretted the "stuffy atmosphere" produced by all that burning gas, which still left "dark corners" where the merchandise could not be seen to best advantage. A year later he installed a set of Brush's powerful arc lights, the system that had just won the endorsement of the city's Franklin Institute. The public flocked to Wanamaker's once again, but this time many held back, warned by rival merchants that the new lamps would soon explode and burn the whole building down. Those who wagered that "Wanamaker's folly" would end up in disaster lost their bet.[19]

A New Orleans levee at night, 1883.

Brush's system allowed Wanamaker to create acres of dazzling retail space under one roof, arranging his glittering showcases through the building's once dark interior and even down in the basement, where unventilated gaslight would have produced little more than a smoldering murk. Throngs of shoppers flocked in, not only to see the artfully illuminated merchandise but also to take in the spectacle of this "wonderful store" itself, which seemed particularly "beautiful and stirring" at night. In similar "grand emporiums" that sprang up in other cities across the country, all followed Wanamaker's lead in harnessing electric light in the service of sales. Boston's Jordan Marsh, for example, installed the equivalent of fifty thousand candles, a sight that local boosters declared to be "the finest window display ever seen in the world."[20]

While shoppers might be helpless to resist the lure of the new department stores' grand electrified spectacle, the technology also empowered them. Long before the age of electricity, the quality of a store's light had been part of the turf on which buyers and sellers struggled for advantage. Storeowners tried to create attractive lighting with the limited tools at their disposal, but they also knew to put their shoddy merchandise in the darker corners. As one guide to successful salesmanship put it, goods belonged where customers could see them, but "just short of where the defects in the goods begin to show. The more perfect the goods are, the stronger can the light be." Gaslight improved on the limited natural light available, but its yellow flicker often misled the eye. Savvy buyers understood that if they wanted to see true colors and the quality of workmanship, they needed to carry goods to the window or even out on the street. The first big retail stores to adopt the electric light acknowledged that the new technology gave customers an edge in this game of hide-and-seek. A store bold enough to install electric light, they suggested, had nothing to hide.[21]

By the mid-nineteenth century, some trades and professions demanded "night work" as a matter of course, serving the market demands of the growing cities and the dictates of new transportation and communications technologies. When the morning trolleys carried most workers to their jobs they also carried home this exhausted and pallid crew of "toilers of the night"—telegraphers, typesetters, and journalists, waiters and musicians, milkmen and bakers, sanitation workers, prostitutes, and policemen. Before electricity drove their profession to extinction, gas lamplighters also joined that group, heading home after early morning rounds extinguishing their lamps.[22]

Other jobs demanded night work only intermittently: stevedores unloading perishable goods, field hands bringing in a harvest and truck farmers bringing crops to market, construction workers pressed to meet a deadline, and mill workers forced to put in late hours when machines broke down. Many urban shops stayed open after dark and during their busy seasons stores required shopgirls to work even longer hours, sometimes standing on throbbing feet till near midnight. Those working under gaslight complained that "after an hour or two your head feels baked and your eyes as if they would pop out." Working under arc lights was cooler, but their blinding glare led some to wear blue-tinted sunglasses, a practice frowned on by their employers. "The only comfort," one saleswoman thought, "is you're with a lot of others and don't feel lonesome." Hotel clerks likewise complained about standing for long hours each night, their eyes bombarded with the flashing glare of the arc light's "electric suns."[23]

A Chicago eye doctor warned about the way that light "played mischief" with workers' eyesight. He told a newspaper reporter about one of his patients, "a man whose eyesight had been nearly ruined by an electric light and whose nerves had been shattered by the same cause." The doctor advised workers to keep away from the arc light, a luxury that those stuck on the night shift could ill afford.[24]

Some workers resisted the late hours made possible by the lighting revolution. An early strike by New England's female millworkers protested in vain against the odious new practice of "lighting up," and throughout the nineteenth century labor unions continued to oppose night work, part of a broader movement to win the eight-hour day. In many industries employers found that they could only lure workers onto the night shift by offering premium wages.[25]

The controversy over night work also played a central role in attempts by labor leaders and middle-class reformers to limit the use of child labor. In the antebellum period, many Americans denounced the use of children in England's mills, factories, and mines, but as the Industrial Revolution expanded into the United States the practice came with it. Some northern states did ban night work for women and children, but those rules were poorly enforced and in the end gave a competitive edge to newer mills in the South, where children enjoyed even less protection until well into the twentieth century. Even when some states banned night work for children in factories, loopholes allowed seasonal industries to employ them with no such safeguards. In Maine, for example, children continued to work at all hours in sardine canneries during the harvest season. They were roused from their beds whenever the boats arrived after dark, and worked through the night cutting fish. An investigator reported that "the cannery whistle, not the sun, brings day to the Maine coast."[26]

As new machines and the longer workday intensified production, children also did night work in bottle and box factories, in coal mines, and in textile mills. The ramshackle villages around southern mills served as a visible reminder that for these workers the new technology offered no comfort or convenience, but was a tool of economic production that only intensified their exploitation. Each morning before dawn, workers left their candles and dim tallow lamps to walk through unlit streets, guided forward by the tall factory windows already ablaze with electric light. To those who did not have to work inside the mill, the sight could be "indeed beautiful." Viewing Atlanta's first mill illuminated with arc light, a reporter noted, "The long lines of windows all night shine out splendidly in the dark. The light casts a weird, strange aspect over the entire landscape around the factory." A young girl forced by her family's poverty to work the night shift saw all this differently. "When I first went to work at night," she told an investigator, "the long standing hurt me very much. My feet burned so that I cried, and my back pained all the time. . . . My eyes hurt me, too, from watching the threads at night. The doctor said they would be ruined if I did not stop the night work. After watching the threads for a long time,

I could see threads everywhere. Sometimes I felt as though the threads were cutting my eyes."[27]

Middle-class folks also found that artificial light intensified the strain of making a living in the modern world, especially for those doing what many called "literary work." Though free from all the other dangers and abuses faced by industrial workers, this class of writers, journalists, and skilled printers faced their own struggles with the exhaustion brought on by the intensified pace of industrial capitalism, racing to meet a growing public thirst for information that could be met only with night work. The problem preceded the electric light, as mid-nineteenth-century writers complained that work pressures were driving them to nervous exhaustion, drunkenness, and an early grave. In the first iteration of what has become a twenty-four-hour news cycle, journalists rushed to produce multiple editions of their paper each day, starting on next morning's edition as soon as the last evening one was "put to bed." When the famed New York editor Horace Greeley died in 1872, his friends recalled his complaint that he "had not had a good night's sleep in fifteen years," an insomnia blamed in part on the physical and mental demands of long evening hours spent working under gaslight. Night work is "killing our literary men," one noted in his eulogy for Greeley. The moral of the great man's premature death was that "the money that a man makes out of midnight toil is paid out in funeral expenses." Better lighting made it easier to work but harder to rest.[28]

Such problems were not confined to literary types. Energetic nineteenth-century Americans liked to boast of their "wide-awake culture," but critics warned that in fact they needed much more sleep. The frantic pursuit of wealth led too many to be "toiling, thinking, planning, scheming all day and far into the night." In the new economic order, enjoying the evening repose that had once seemed mandated by God and Nature now felt like an "unpardonable sin." The results of these "late hours and early hours, travelling and reveling" were evident, as urban Americans struggled with a plague of nervous exhaustion, drug and alcohol addiction, "softening of the brain," baldness, spectacle wearing, premature aging, and early death.

The long hours working under gas lamps did not bear all the blame, of course. Medical men also preached jeremiads against tobacco, sleeping in overheated and unventilated rooms, the jarring interruptions of street noise and telephones, and the enervation of body and soul caused by a host of other modern conveniences and inconveniences. But experts agreed that many of the mental and physical ills suffered by American city dwellers could be traced back to gaslight's interference with the natural order of day and night, work and rest. How much worse would things become, they wondered, now that Edison had unleashed a new invention that meant "the day would practically have no end"? "The nervous energy of the race," as one put it, was about to be "called upon to meet a new strain."[29]

Gilded Age Americans were both fascinated and troubled by the new machines that so rapidly transformed their work lives, engines of enormous production that seemed to create abundance and misery in equal measure. No less than the new looms, furnaces, and steam engines, electric lighting systems played a central role in this economic revolution. As much as any other machine, electric light contributed to the intensification of production in the late-nineteenth-century Industrial Revolution, and the creation of a very modern sense of perpetual urgency.

And yet except for the battle over child labor and night work, critics of these changes did not often single out the electric light as a prime culprit. Many squinted in the glare, but few longed to return to the yellow flicker and noxious fumes of gaslight, a technology

A Brooklyn power station.

that quickly seemed as quaint and outmoded as the ancient taper. In one industry after another the light arrived, forcing workers to adjust their eyes and their work habits accordingly—but then the light receded into the background of their consciousness. Even as it created new ways of working and changed the entire rhythm of the day, it became harder to see. As one observer at the time put it, "Whatever constantly enters into the daily life soon becomes an unnoticed part of it." Increasingly commonplace in offices, factories, and depots, electric light receded into the background, its power undiminished by its invisibility.[30]

At the same time, the technology became a visible and essential part of the way urban Americans enjoyed their leisure time, playing a central role in the creation of the "nightlife" that became a defining characteristic of modern cities—a new form of play that complemented the culture's new forms of work. The same technology that accelerated production and put so much stress on Americans' eyes and nerves also promised them relief—countless variations of electrified fun that made bright lights an essential ingredient in modern Americans' pursuit of happiness.

Five

Leisure Light

Throughout history, the wealthy always enjoyed the privilege of artificial light, able to spend lavish sums to burn candles or oil late into the night and to hire the servants who worked long hours the next day cleaning and trimming the lamps. In the preindustrial age, farmers and artisans used light much more sparingly, and most often not for leisure hours but as a necessary tool of production, a way to extend working hours, especially in the short winter days. To avoid the expense of candles, many toiled as long as possible in the gathering dark, a practice some called "keeping the blind man's holiday." The spread of gas for home lighting in the mid-nineteenth century began the slow but steady democratization of light, turning an evening of illumination from a luxury into a middle-class comfort, at least for those living within the better neighborhoods in the urban core. And in shops, theaters, and parks, gaslight enriched cultural life, pressing back the dark to stake out new time and space for an urban sociability that was enjoyed by rich and poor alike. "Sundown no longer emptied the promenade," as

Robert Louis Stevenson put it, "and the day was lengthened out to every man's fancy."[1]

Now the arrival of electric light furthered this trend, offering more affluent consumers an abundant light for their homes and providing many more with an increasingly vibrant and varied menu of evening entertainments. A decade after Edison unveiled his first incandescent system in downtown Manhattan, American city dwellers amazed themselves at the late hours they began to keep, losing what *Harper's Weekly* called "any prejudice that may have heretofore obtained in favor of the garish day." In fact, one Chicago journalist scoffed at the bromide that those going "early to bed" would turn out healthier, wealthier, and wiser. Such "ethical absurdities" were "fast going out," he wrote, especially for the young. "The old idea of going to bed with the chickens and rising with the larks grew out of the fact that in old times the opportunities for pleasure at nights were not as great as they are now."[2]

Better street lighting made night travel safer and guided pleasure seekers to a growing number of entertainment venues, each one beckoning with a blaze of light. The transformation of evening nightlife began under gaslight, which was adopted by most theaters but came with some nasty side effects. Much brighter than candles and oil lamps, the gas burners also consumed far more oxygen, replacing it with what one unhappy theater patron described as a "carcinogenic fog." Under gaslight, the pleasures of the theater were tempered by this exposure to poisoned air and oppressive heat. After one too many enervating evenings, waking up the next morning with a gas-fume hangover, some swore off evening amusements altogether.[3]

Fire proved an even greater danger, as open gas flames sometimes ignited stage curtains and costumes. Public alarm reached fever pitch in 1887 when a gas fire burned a Paris opera house to the ground. Two hundred people were killed, including many actors and dancers caught in the flames and others trampled in the mass panic that erupted when the gaslights blew out. Only months later a gas fire gutted a brand-new theater in Exeter, England. While the actors tried to restore order, the gas mains exploded, instantly filling the theater with smoke and flame. Most on the ground floor escaped after suffering a "dreadful crushing," but more than a hundred in the gallery died, trampled or "roasted" in a narrow stairway.

In the wake of these sensational disasters, civic leaders on both sides of the Atlantic called for new regulations to ensure public safety—more and wider exits, fireproof curtains and paint, and most importantly the replacement of gas with incandescent light. The city council in Paris gave all theaters, cafés, and concert halls three months to replace their gaslights with electric lights. American governments imposed no such restrictions, but managers of the better theaters moved quickly to install electric systems and advertised this safety feature to their customers.

The new light not only made theaters cleaner and less dangerous but also helped to make them bigger. The power and efficiency of electric light made possible the large entertainment emporiums that cities and private investors created in the late nineteenth century, where they fed a growing hunger for sport and spectacle. P. T. Barnum was among the earliest to incorporate the light into his entertainment extravaganzas, using it under the enormous canvas tent of his traveling circus and in the "grand room" of Madison Square Garden. There many thousands at a time flocked to see chariot races, walking matches, archery and shooting contests, dog shows, steeplechases, and "other entertainments requiring much space." When the Garden was rebuilt in 1890, making it the world's largest public entertainment hall, four thousand "tiny balls of electric light" illuminated the cavernous structure, which was capped by a tower topped with Augustus Saint-Gaudens's lovely but controversial statue of the goddess Diana, her nude golden body illuminated by a crown of incandescent bulbs. The entire hall became a "Temple of Electricity," as one put it. "Whatever else is furnished to amuse or interest, the exquisite lighting effects are ever in sight, so that no one can feel uncomfortable or bored."[4]

While audiences enjoyed the cooler, cleaner air provided by electric lights, the performers at first had their doubts. In early attempts to use arcs, actors objected not only to the disorienting glare but also to the light's power to expose a performer's every blemish. Makeup that looked fine when put on under the dressing room gaslights turned cartoonish in the cold blue electric light, betraying what one actor called "too many secrets of texture and complexion." In at least one case, actors refused to let the show go on, boycotting the new light. A similar problem affected the old stage sets, which looked convincing in the flickering yellow haze of gas but crudely artificial under stronger light. Over time directors learned how to temper the arc and adjust makeup and set designs, abandoning the use of set paintings in favor of more realistic three-dimensional sets, painted in a palette better suited to electric light. And when incandescent bulbs arrived, theater artists quickly embraced their possibilities. On both sides of the Atlantic, audiences enjoyed glowing moons, the flash of artificial lightning, and blizzards of incandescent light that "drove paper snow from the stage." Unlike gas and arc lights, the new bulbs could be arranged in any position and tinted any color. Empowered by the new technology, late-nineteenth-century stage artists developed an increasingly sophisticated palette of light, painting the stage with it to create the manipulations of attention and mood that have become a staple of the theatrical experience.[5]

At times the actors themselves sported the lights—in one production, the devil's horns glowed red with incandescent lamps, and Edison joined the act in a New York ballet, rigging tiny battery-powered lights that twinkled on the forehead of each ballerina, an

Alice Vanderbilt as "The Spirit of Electricity," 1883.

effect that some found "dazzling." Another troupe installed electrified metal plates on the stage; when dancers crossed them, the current formed a connection through wires in their shoes, setting alight the incandescent bulbs in their costumes. Perhaps inspired by these spectacles, some of London's "dashing demi-mondes" began to wear electrified diadems and brooches, the small batteries hidden in the folds of their dresses. Thanks to the novelty of incandescence, "a pennyworth of glass" shined bright enough to "eclipse a duchess' diamonds or rubies," producing a "shocking effect" when it first appeared. For a time, cheap electric brooches and pins became the latest thing, though Mrs. Alice Vanderbilt soon trumped all rivals by wearing a diamond-studded electric light costume to her family's annual ball, one of the most talked-about social events of the New York season.

While light was a powerful tool for stage performers, and a bauble for London prostitutes and the wives of robber barons, early pioneers in the field of illumination engineering embraced electric light as a medium all to itself. Some customers balked at the price of electric light and others feared its safety, but none doubted its beauty. Building on this appeal as the best possible advertising, the lighting companies staged increasingly elaborate spectacles. Edison's team set the standard as early as the 1881 Paris exposition. When the United States hosted its own International Electrical Exhibition in Philadelphia in 1884, the hall glowed with the electrical equivalent of a million candles. While exhibitors competed to attract visitors to their displays of "electricity's wonders," all eyes were drawn to the exhibition's grand centerpiece, an illuminated fountain that looked like "ever changing sheets of parti-colored liquid fire."[6]

Through the late nineteenth century, electric companies partnered with public officials in the staging of world's fairs and expositions, using them to sell the public on electricity by showcasing the illuminator's art. At most expositions the electrical industry enjoyed an exhibition hall all its own, but the grandest advertisement was the fairgrounds itself, especially for visitors after dark. In the early 1880s, when the technology was quite new, the expositions could simply rely on the novelty of the new technology to draw a crowd. Visitors swarmed in for a first glimpse at the glare of naked arc lights and a chance to see the technology up close. "The electric light machinery attracts everybody," noted one visitor to an 1883 exposition in Wisconsin. "This great invention naturally interests the educated and the ignorant alike."[7]

Each of the great world's fairs strove to surpass all others in offering the most eye-boggling electric spectacles. Cincinnati's centennial celebration in 1888, for example, promised a "fairy fountain whose splendor has never been duplicated," and street illuminations "of unrivalled brilliancy and profuseness." Edison's sixty-five-foot-tall light tower, first unveiled in Paris, drew crowds to the Minneapolis Exposition in 1890, which also included a reflecting pool full of colored light bulbs and goldfish. Designed and operated by the new electric companies, these displays delighted masses of visitors while delivering a powerful message about industrial capitalism's command over Nature and the central role that electricity would play in the nation's quest for a better future. Historian David Nye suggests that these expositions allowed millions to envision an electrified city of the future, a man-made environment that was orderly, elegant, and saturated with light.[8]

For Chicago's 1893 Columbian Exposition, the United States celebrated itself with the grandest light show yet. Westinghouse installed the largest electric power plant the world had ever seen, more light condensed on the fairgrounds than in the entire city. Thousands of incandescent bulbs etched the arches and rooflines of the fair's temporary but impressive Beaux-Arts buildings as if "tipped with the sparkle of myriad diamonds." From a distance their reflections danced in the waters of the Great Basin, multiplying the dazzle,

The Edison Company's light tower became a central attraction at many expositions.

. . . .

especially when the fountains turned on each night. Fifty feet high, these "masses of illuminated water" provided a shimmering canvas on which electricians played the "princely nineteenth century trick of glorifying water with color, and transforming it into the semblance of precious gems . . . somersaulting like explosions of amethysts or pearls or emeralds, turquoises or sapphires." Bringing focus to this disorienting riot of colored light, a powerful spotlight picked out statues and dramatic architectural vistas, plucking them "out of the nightlights into the full splendor of broad noon . . . like a bit of ivory on black velvet."[9]

As historians have often pointed out, fairgoers gazed in awe at the neoclassical buildings, boulevards, and fountains of the Chicago fair's White City, but they flocked to the more eclectic, lowbrow amusements of the midway. Inspired more by Barnum than by Beaux-Arts, a new breed of entertainment entrepreneurs offered a different form of cultural sensation, what historian John Kasson has called "a hurly burly of exotic attractions." While it lacked the candlepower of the majestic White City, the fair's midway did not ignore the electrical excitement but offered a world where the light lured without driving out all shadow, leaving room for attractions like the "Persian Palace of Eros." Rising high above both the midway and the stately White City, Ferris's "great wheel" proved to be the fair's most popular attraction, glowing each night with over two thousand bulbs "like a circle of fire."[10]

Inspired by the success of electrical spectacle at the nation's major expositions, entertainment venues raced to add more light. One magazine satirized a hotelkeeper at a seaside resort who expected to "barrel money this summer," since he had installed "an electric light on the lawn near the fountain, and another on the shirt-front of the head clerk." But few matched the grand schemes and ambitions of Erastus Wiman, a real estate developer who staked his fortune on the lure of electrified entertainment. In the late 1880s, Staten Island remained distant from the lives of most New Yorkers, but Wiman bought up acreage, along with rail and ferry service, hoping to cash in by turning the borough into "the greatest suburb of the greatest city in the greatest country." He planned to lure New Yorkers to Staten Island by offering electrified fun at a modest price. On its shores he constructed a massive amusement complex complete with a perpetual circus, a cricket field, a ballpark for his own "Metropolitans" baseball team that sometimes featured night games, and a series of epic summer extravaganzas. In its inaugural year he hosted Buffalo Bill's Wild West Show, a tribute to the fast-disappearing frontier that nicely combined America's mythical past with its modern-day fascination with glittering spectacle. Gleaming under the beams of hundreds of arc lights and swept by nine enormous spotlights, the troupe of 250 cowboys and Indians roped, shot, wrestled, and wrangled a herd of buffalo, cattle, and elk. The show probably consumed more candlepower in an evening than all the candles and kerosene lamps in the Dakotas could generate in a year.[11]

On summer evenings, New Yorkers escaped the city heat by buying a ticket on Wiman's ferry. While crossing to Staten Island they enjoyed the cool harbor breezes, gazed at the city's increasingly electrified skyline, and got a close look at the new Statue of Liberty, its torch emblazoned by a powerful arc lamp. But those sights only served as an appetizer for Wiman's customers, who found on the island's "gay shores" a glowing fairyland of fun, a place to embrace what he called "the gospel of relaxation." After a promenade on the lawn, festooned with strings of colored light like a "glistening girdle of gems," the visitors enjoyed a band concert capped with a cornet solo. But the evening's highlight was the show put on by Wiman's massive illuminated fountain. Twenty-five men ran it, operating fifteen jets of water that shot 150 feet in the air, each colored by the rotating flashes of a thousand tinted Edison lamps and played on by powerful beams from an arc spotlight. As a *New York Times* reporter wrote, "the water was seen under advantages to which the fluid is rarely subjected." Geysers that looked like "molten silver" melted into "blood red jets, falling in purple spray." The fiery column could be seen all the way across the harbor, and beckoned New Yorkers strolling along Battery Park. Americans had seen this sort of extravagant light show at grand public exhibitions, and London and Paris had incorporated them into some of their most important public parks. Now the electrified fountain entered the field of commercial amusement, a dazzle one could enjoy any night for a modest ticket price. In the grand style of an American showman, Wiman guaranteed that his fountain was the world's biggest, and more beautiful than any found in Europe.[12]

Through the summer of 1887, Wiman's ferries carried ten to twenty thousand paying customers each night to see not only his fountain, circus, and baseball team but also the biggest theatrical event of the season, *The Fall of Babylon*. This was no plebeian fireworks display, Wiman insisted, but a "spectacular and historical drama." Fifteen hundred actors and dancers, dressed in elaborate costumes designed in Paris, cavorted on a massive set that blazed with "the most powerful electric light known to science." In a touch of drama that even Barnum could not have arranged, just as three hundred carpenters put the final touches on the stage set, a lightning bolt smashed its "Tower of Babel" to splinters. The show went on, catering all summer to "the rich and the poor, the good and the bad"—anyone who had a ferry ticket and the twenty-five-cent admission fee. The better sort claimed that they came not for the sensation but to be edified by this illuminated history lesson. Wiman boasted that no fewer than a thousand clergymen had seen and endorsed his show. Others came for the catapulted fireballs, flaming castles, flying arrows, and chanting Chaldean priests. As a reviewer in the *Times* summed up the show's enormous appeal, "The fall of anything, especially of anything so rich in suggestion as Babylon, will ever be fascinating to the dwellers on this planet. Best of all, it is an easy thing to get to Babylon. The fall is over as early as 9:30, which makes it possible to get back to the city by 10 P.M." The show's only critics may have been the Mets baseball players, whose field adjoined the stage set; all season long, any hit that landed inside Babylon was ruled a ground-rule single.[13]

Colonel Hickenlooper, the mighty gas man, brought his family from Ohio to see the spectacle and noted in his diary that the main attraction was "nearly four hundred scantily attired young misses trained to dance very artistically." He remained silent about the electric light display, a thousand Edison arc and incandescent lights bathing a six-acre stage, as big as a city block. The friends of electricity did take notice, of course. Hailing *The Fall* as the greatest piece of "spectacular art" that New York had ever seen, they pronounced it "largely a triumph of electric light."[14]

In spite of the show's popularity and an even greater success the next summer when Wiman staged Nero's burning of Rome, his scheme to build Staten Island on the lure of electrical entertainment soon failed. His power station burned down in 1891, and during the financial panic of 1893 he was forced to sell the Mets and then his amusement park. Not long after doing a stint in prison for check forgery, he died penniless. Only his marvelous fountain survived, moved to a new location in Chicago's Lincoln Park, where crowds of twenty to thirty thousand enjoyed the spectacle on summer nights.[15]

In spite of Wiman's sad ending, other entrepreneurs continued to build on his attempt to market the electrified spectacle as a commercial entertainment, a trend culminating a few years later in the transformation of Coney Island into a luminous "city of pleasure." A strip of sand on the far edge of the borough of Brooklyn, Coney

Island erected its first arc light in 1879. In these years cramped city dwellers came to enjoy the ocean breezes. The arc light even tempted some to try night swimming, an "electric revel" that the *Times* thought "lunatic." But the place was best known as a haunt for thieves and gamblers, a spot where "no respectable person would care to be seen." All that changed when some of the era's greatest entertainment entrepreneurs built massive amusement parks on that strand, eventually burning so many bulbs that ships could see Coney Island from four miles out at sea, making it the first sight of the New World for many immigrants. No longer a "social sore," Coney Island became "the most extensive and best show place in the world."[16]

During the warmer months the resort welcomed large crowds each day, but many liked the place best at night, when it became "a realm of fairy romance in colored light." On summer evenings tens of thousands paid a modest admission price to enter one of the three great amusement parks. They thrilled to watch reenactments of the Fall of Pompeii or the Johnstown Flood, toured imitation coal mines and "the sewers of the great cities," rode on slides, swings, Ferris wheels, and loop-the-loops. But the greatest attraction of all, many noted, was the sight of everyone else, the "great mixed mass of humanity" all enjoying themselves like children. This easy democratic interchange was artfully orchestrated by the parks' designers, who created ample opportunities to throw people in each other's arms, or at least to the center of each other's attention—spiraling chutes that tumbled pedestrians, funhouse mirrors, jets of compressed air that raised unsuspecting women's skirts, and rides like the Barrel of Love.[17]

Critics derided the amusement park as "an artificial distraction for an artificial life," the worst excess of a new form of mass-market commercial entertainment. But Coney Island's success confirmed that Americans would line up in great numbers to enjoy cheap sensations and childlike camaraderie. It also confirmed electric light's

Luna Park at night, circa 1905.

value as a tool for creating a euphoric and trusting mood that so many found irresistible. The millions of light bulbs that blazed over Coney Island each night offered an attraction all their own, stimulating the senses to a fever pitch and making everyone and everything ultra-visible. At the same time, they provided a reassuring sense of safety that made it easier for customers to let go and enjoy an experience that was in all other ways carefully designed to feel disorienting and chaotic. Enjoying temporary liberation from some of the confining boxes of class, gender, and ethnicity, Americans from many different neighborhoods and walks of life mingled without fear or suspicion. Just as city policemen erected powerful lights to drive away criminals, the brilliant lights of Coney Island made crime and danger seem unthinkable. As one visitor put it, "any one who can rob or even practice the mildest deception under the present white light of publicity is deserving of the swag."[18]

The greatest of these entertainment emporiums, Luna Park, turned the dynamo itself into an exhibit, a nod to education that reflected the amusement park's debt to the electrical expositions that pioneered the art of the illuminated spectacle. The whirring machine gleamed in white enamel and gold paint, supervised by an electrician decked out in white gloves and a brass-buttoned uniform. To all who would listen, he stood ready to give an entertaining lecture on the mysterious pulsing force that powered all of the park's amusements.

Across the country, electric companies took a keen interest in Coney Island's success. The park produced weekend and evening customers for electric trolley service and created demand for electric light on summer evenings, when long hours of daylight usually hurt business. Soon the owners of traction and utility companies in many other cities invested in their own "miniature Coney Islands," each using the lure of abundant, colorful light to create the feeling of safe family fun. "The American people are always eager to be amused," noted the *Electrical Age*, "and men of foresight and enterprise have long since realized that fortunes were to be made catering to this pleasure-loving instinct. But it cannot be done on a large or successful scale without the aid of electric light."[19]

Within a few short years, the same light first unveiled to great applause in major urban parks and at public exhibitions found its way not only into the grand new emporiums of amusement but all the way down to "fifth-rate barber saloons, cheap eating houses, outside peanut stands and in modest lager beer saloons, where the 'masses' are seen and do congregate." Arcades and movie theaters

"Surf-Bathing by Electric Light at Coney Island," 1904.

lured nickel-toting customers with "fantastic constellations of colored lights," while one of the most brilliantly lit districts in the world could be found in New York's working-class Bowery.[20]

Beyond these entertainment venues, the new public lighting systems nurtured a rich urban nightlife, as neighbors felt safer on their stoops and sidewalks long into the evening, young boys played stickball under streetlamps, buskers, evangelists, and street vendors drew evening crowds, and parks and boulevards offered extended hours for courting and promenading. In winter, urban skating parks stayed open late, and as one put it, "no toboggan slide is considered complete without electric lights." In Atlanta young ladies cycled under the electric lights on Peachtree Street, hiring a "professor" who taught them "the art of equilibrium on a wheel." In St. Louis the light shone down on a "massive craps game," bright enough to give most of the gamblers ample warning when the police tried to grab them. And in Los Angeles temperance reformers used a pair of electric lights to draw a crowd for their soapbox crusade, though the audience mostly heckled the "sacred cause" and seemed to one reporter to be "urgently in need of prohibition in the most violent form."[21]

World's fairs, amusement parks, and entertainment districts provided a laboratory for the invention of electrified fun, creating an aesthetic that suffused every other part of the emerging culture of urban nightlife. Only a few months after Edison's Menlo Park demonstration, some Boston baseball fans were eager to apply the new technology to the American game. On a September evening in 1880, teams from two city department stores faced off under clusters of incandescent light on Nantasket Beach. While some spectators thought the light was "very strong and yet very pleasant," the players compared it to playing under moonlight. Three years later,

Electrified nightlife on the Bowery, 1891.

the new Jenney Electric Company in Fort Wayne, Indiana, hoped to drum up publicity by staging a game under its much stronger arc lights, rigging twenty of them around a field and drawing a couple thousand spectators for a game between a professional team and the local Methodist college. Boosters mistakenly claimed the honor of being the first town to play baseball under "the rays of artificial sun," a "historic" feat they predicted would cause Fort Wayne's name "to be mentioned wherever civilization extends." Instead, most considered the experiment a "dead failure." A New York sportswriter sarcastically suggested that the Polo Grounds should follow Fort Wayne's example, since fans would rush to see their favorite players "sweeping around over the grounds with lanterns in their hands, looking for the ball."[22]

During the late nineteenth century, public interest in baseball grew enormously. Fans were ready to spend money at the ballpark, but games could be played only by the light of day, enjoyed by those one historian described as "the well-to-do, the unemployed, or school kids." Team owners recognized the profits to be made if they could offer baseball in the cool evening hours when most were off work. Once someone figured out how to properly electrify baseball, a sportswriter conjectured in 1893, "the possibilities of the game's earning could hardly be estimated." Across the country, teams looking for an extra take at the gate offered night games as a novelty, some even experimenting with the use of a phosphorescent ball. But most fans declared night baseball a "farce." Pitchers had to ease up to give batters a chance to see the ball, and fielders struggled against both shadow and glare, a problem even Edison tried but failed to remedy. After Hartford, Connecticut, tried night baseball for the first time, the local paper reported, "Only a few innings were played, and nobody knows the score." Night baseball remained a minor-league novelty until the technology improved in the 1930s.[23]

When some suggested running horse races under the lights, the conservative guardians of the "sport of kings" scoffed at the idea. "It would be difficult to conceive of a more absurd and useless practice," a leader of a New York turf club declared in 1881, when electric light was still in its infancy. "Why should we run horses at night, when we have daylight at our disposal?" A decade later, a St. Louis track made what it claimed to be the first successful experiment, rigging dense clusters of incandescent bulbs along the track and saturating the turns with spotlights. Spectators came to the first trials just to see if the thing could be done, and left raving about a sight that many found "really beautiful." The illuminated track made a "striking picture," as one put it, while the gaudy colors of the jockeys' silks seemed even more vivid under the lamps. And the final turn of each race looked more exciting than ever, as horses emerged out of the far shadows and streaked in front of the brilliantly illuminated grandstand. In this heightened visual field even the jockeys' expressions seemed quite visible as they barreled across the finish line.[24]

Similar experiments were tried for most sports, and early witnesses of these spectacles responded in similar ways. Watching a sporting event under the lights, they found, was not just a shadowy copy of the same thing by day. The powerful lights and contrasting shadows lent familiar scenes a new look, a vivid theatricality. As the technology improved in the early twentieth century, many declared a preference for the sensations of watching a night game or race, a surprising revelation at the time but one that still rings true for many long after the novelty of night games has worn off.

Playful displays of electric light marked out a new urban landscape of leisure, the amusement parks and stadiums, theater and commercial districts zoned for pleasure. These brilliant corners of American life were only the most visible showcases of a new aesthetic that spread across the culture in the late nineteenth century. In a similar way, the light entwined with Americans' idea of the holidays, an important visual dimension of those few times in the calendar year when people invited themselves to stop working and have fun. The German tradition of lighting wax tapers on an evergreen tree had become a staple of Victorian Christmas celebrations on both sides of the Atlantic, a quaint but dangerous practice that led many families to admire their tree, but only with blankets and buckets of water at the ready.[25]

The Edison Company's great light designer, Luther Stieringer, quickly saw the possibility of using incandescent bulbs to turn the Yule tree into a safer but gaudier extravagance. At a Boston fair in 1884 he frosted a forty-five-foot tree with over three hundred colored lights, operated on twenty-four different circuits, that could be turned on and off in a variety of combinations, often in time with music. The "strangely beautiful" effect struck observers as unique, though it would not remain so for long. Each new holiday season upped the ante, and within a few years lighting designers added more elaborate and flashing bulbs, revolving tree stands, and the cold but convincing flicker of an artificial fire at the hearthside. "The illusion was perfect," one observer thought, "and the cheerfulness of the room was not in the least diminished by the exposure of the trick." Just as it had done for Barnum's shows and Coney Island, bright lights soon marked Christmas as a space and time apart, where families could share a childlike euphoria, a lightness of spirit induced by luminous color. Merchants soon recognized that they could use light to provide the same experience to shoppers, drawing them in with elaborate window displays and festooning their stores with enough colored light to confuse their customers over the difference between celebrating and shopping.[26]

Electric light never managed to replace fireworks as America's flashing light of choice on the Fourth of July, perhaps because of that holiday's association with the nation's preindustrial past. But by the early twentieth century, Americans began to incorporate electric fun into New Year's Eve, adding it to the old standbys of alcohol, confetti, and horn tooting. The Times Square electric ball, studded with 125 Edison lamps, descended for the first time in the last seconds of 1906, but well before that revelers celebrated with spotlights, Japanese lanterns, and other electrical displays. In fact, downtowns blazed with so much light that the moment of midnight had to be marked by turning the lights *off* for a few seconds, then on again at the crucial moment. Swanky restaurants added some spice to holiday dining by preparing something special for their customers to see when the lights came back on—an unfurling American flag, or a young woman in a Cupid costume tossing flowers from a cornucopia.[27]

The light also played an increasingly central role in political campaigns, those carnivals of democracy that roused the passions of so many Americans in the Gilded Age. When James G. Blaine ran for president in 1884, New Yorkers turned out for one of the largest parades in the city's history. Long after dark, thousands jammed the sidewalks, a spectacle that the *New York Tribune* called "pleasing to the eye and inspiriting to the soul." Bands and marchers from around the country carried the usual torches, colored lanterns, and spark-belching roman candles, but all commentators agreed that among this "succession of splendors," the highlight was a marching corps from the Edison Company, each man wearing a glowing electric light helmet. Led by the great inventor himself, they towed along a steam-powered dynamo, linked to each helmet through an electric wire each man carried on his shoulder. As they passed the reviewing stand, their helmet bulbs twinkled while their steam engine did double duty, running a calliope that serenaded their candidate with "Hail the Conquering Hero." The publicity probably helped Edison more than it did Blaine, who lost the election to Grover Cleveland a few days later. From then on, electric light's power to command attention and fire enthusiasm made it an important tool in campaigns. By the time William Jennings Bryan made his spectacular runs for the presidency at the turn of the century, his major speaking engagements included a flood of spotlights, massive illuminated portraits, and his name spelled out in incandescent bulbs.[28]

The Edison Company's "Electric Torchlight Procession," 1884.

Light's stimulating power could not be denied, but some feared that it was fast becoming addictive. Middle-class temperance reformers worried that working-class men were too often lured away from their homes and families, tempted by the saloon's heady brew of cheerful light and strong drink. After an exhausting day of work, who could bear to stay home at night with his family in a cramped and gloomy apartment, lit perhaps by a kerosene lamp or a smoldering tallow wick, when the friendly glow of the saloon beckoned? And so urban reformers pushed for a further democratization of light—only when working families enjoyed comfortable, brightly lit homes, safely illuminated parks for recreation, and "well-lighted reading and club-rooms for working-men" could society hope to make any headway against the corrupting effects of the liquor trade. Reformers applauded when grand public institutions such as New York's Metropolitan Museum of Art installed their first lights, expanding their mission of cultural uplift to reach more workers who toiled during the day.[29]

The lure of urban America's electrified nightlife contributed to another alarming trend in the late nineteenth century, the steady stream of rural men and women who abandoned their farms and small villages to seek new lives in the city. Many came looking for economic opportunities, but plenty came not to strike it rich but to find richer social lives. "The village is dull," as one put it, "not only to the man pursuing light amusements, but to him who seeks cultivated associations, for in these days the cities are the centers of intellect as of wealth." Everyone knew that Gilded Age cities were overcrowded, polluted, and corrupt. But in spite of all this, they managed to offer a growing number of services rarely or never found in the country—not only an expanding range of evening entertainments but also great libraries, lecture halls, art museums, and civic groups. "The desire for amusements" exerted some influence, as one observer put it, but migrants from the country also sought "better social, educational and religious advantages."[30]

This trend jarred against the nation's long-cherished belief that the fate of the Republic rested with its independent, landholding farmers. Echoing this Jeffersonian tradition, pundits in the pulpit and the lecture hall continued to denounce cities as a hazard to the nation's moral and physical health. As Josiah Strong put it, the fast-growing industrial cities posed "a serious menace to our civilization." Another reformer, Charles Loring Brace, insisted that the best way to help the city's struggling masses of poor children was to relocate them to the country. Liberated from the "dark places of the metropolis," orphans adopted by "good country families" could be transformed into responsible citizens, thanks to the healing balm of sunlight and fresh air, healthful exercise, and ennobling physical toil. This worked for some, but social workers found to their surprise that many ghetto dwellers had no interest in moving to the country, while each year many thousands of rural men and women flowed in the other direction. As one expert on urban problems put it, "All efforts to arrest the progress of the cities, and to check the population that continually flows into them, must be fruitless. The great social movements of the age cannot be stopped."[31]

Farms were abandoned and villages depopulated, not just in the drought-ridden West and the impoverished South but across once prosperous areas of New England and the Midwest. Many of the rural migrants were not poor and desperate; rather, they were the more prosperous and ambitious, the young and hopeful. Some commentators blamed this rural depopulation trend on the media. Thanks to powerful new printing presses and railroad networks, urban newspapers and magazines expanded their reach far into the countryside. In their pages, the sons and daughters of farmers found themselves ridiculed for being "hayseeds and country bumpkins," but they also caught a "glimpse of a more attractive life" waiting for them in the city—fancy balls that began at midnight, ball games and amusement parks, luxurious hotels and boulevards blazing with light all night long, and restaurants packed with diners long after farmers and country villagers had gone to bed. The light beckoned many rural Americans, who followed it seeking the pleasures and opportunities of this more exciting modern world. One young woman who migrated to the city in these years summed up her feelings this way— "If I were offered a deed of the best farm I ever saw, on condition of going back to the country to live, I would not take it. I would rather face starvation in town."

America's shift of cultural authority from rural "heartland" to urban center had many causes, and the lure of the cities' electrified nightlife was just one of them. But perhaps more than any other, the

Strollers enjoy an illuminated evening in
New York's Madison Square, 1890.

electric light served as a visible marker of this change, the reach of the lamps' beams tracing the boundary between America's rural past and its urban future. Some rural Americans rushed to town in order to enjoy all the pleasures of an electrified culture. Others came hoping to help build it. Following in Edison's footsteps, these men and women felt sure they had great ideas for inventions that would soon surprise and delight the world and, as important, maybe even make them rich.

Six

Inventive Nation

In 1884, Philadelphia's Franklin Institute organized the nation's first grand exhibition of electricity, an American counterpart to the Paris exposition three years earlier. Other European capitals had already followed France's lead in hosting these technological fairs, and each one confirmed the lead that Americans were taking in the new field of electric lighting. The nation's growing inventive powers surprised and alarmed many European observers, who watched their own cities being lit by American firms and their markets being flooded with a whole range of "new ideas and new applications" developed in Yankee workshops and laboratories. Now the Franklin Institute would invite the world to Philadelphia in order to survey the state of the art and to acknowledge the important role that American inventors were playing in shaping the world's electrical future.[1]

The Franklin Institute had been founded in 1824 to encourage fruitful exchange between artisans and scientists. Modeled on similar scientific societies in Europe, the institute offered lectures and courses, published a journal, ran a library, and encouraged innovations in the practical arts by awarding medals to "worthy workmen" for their new inventions. Among them was Charles Brush, who won a Franklin medal in 1878 for his improved dynamo, a valuable recognition for the young inventor. The Franklin Institute had long sponsored exhibitions in every field of "arts and mechanics" but the 1884 exposition was its first "entirely electrical in character."[2]

Over three hundred thousand visited the exhibition that fall, touring a large Gothic-spired hall constructed just for the occasion. They enjoyed exhibitions of wire insulation, conduits, railroad switches, electrical medical apparatus, alarms, and clocks. They saw new electric motors applied to sewing machines and pipe organs, and an egg incubator warmed by an electric current. "Such an aggregation of electrical devices—experimental, useful and ornamental—has never before been seen in America," a promoter explained. "And vast as it is, it may be considered but the threshold of the coming electrical era." Twice each evening the hall was darkened for a display of the illuminated fountain, its streams of colored water colliding in a prismatic spray that was "thoroughly appreciated by the ladies."[3]

Marking just how quickly the industry had developed, a special exhibit on the history of electricity included dynamos and batteries that had become, in just a few short years, curious relics of the past. Only seven years had passed since Brush first lit up a Cleveland park, and two since Edison's first central station in New York. So much that had once seemed "visionary" had become indispensable to modern life, and none could see an end to this electrical revolution.

The exhibition played an evangelizing role, making special efforts to reach schoolchildren. Not all appreciated the gesture, as the children took loads of company brochures that they had no inten-

The 1884 International Electrical Exhibition, the first of many American showcases of the new technology.

tion of reading, played tag in the aisles, and made unauthorized tugs on steam whistles; one even shorted out a dynamo by dropping a nail into it. Faced with a mostly incomprehensible display of wires, switches, and storage batteries, these children enjoyed the electrical future in their own way, but they seemed to share in the consensus that the exhibition was "the sensation of the day at Philadelphia." A reporter for a Catholic journal went much further. After providing a detailed account of the show's marvels, he concluded that "we may look up to God and allow our hearts to be filled with unspeakable hope."[4]

Few European companies accepted the invitation to this exhibition, expecting to find no customers in an American market dominated by its own inventors and protected by high tariff barriers. European electricians did attend the event's congress of scientists and inventors, and were struck by how rapidly electric light had been embraced in America's eastern cities. "Electric light is flourishing in America," a leading British scientist reported, "much more than at home." Exaggerating its penetration into the American market, many European visitors concluded that only in the United States had electric light become "the rule rather than the exception."[5]

While the "international" exhibition proved to be largely an American affair, eight different lighting companies used this chance to showcase the technologies that were finding eager customers across the country and around the world. One of Edison's rivals, the Weston Company, drew visitors to an "artificial grotto," its perpetual waterfall sprayed with colored light. Charles Brush not only demonstrated his arc lamps but invited visitors into a model drawing room lit by the incandescent lamps invented by his new partner, the British inventor Joseph Swan. But as usual, Edison's display topped them all. His own model room featured artfully illuminated bouquets of flowers attended by a guard wearing one of Edison's electrified helmets. And in the center of the exhibit hall he erected the "Edison Pyramid," a tower studded with twenty-six hundred red, white, and blue lights that his electricians manipulated in a variety of patterns, swirling up and down the structure. Today a similar display might be found on any carnival midway, but it dropped jaws in 1884.

At the closing ceremonies in November, exhibitors formed a parade line around the hall, calling themselves the "Mystic Electrical Order of Kazoos." As they packed up their displays, one of the exhibition's main events began. The judges in previous exhibitions had awarded prizes based on what could only have been a cursory inspection of the equipment, made during the few hectic weeks of the show. The Franklin Institute determined instead to provide the first objective consumer research in the field of electric lighting, conducting an independent test of the competing claims made by Edison and his rivals. Looking past the glitter and glare of the exhibition's electrified waterfalls and pyramids, the institute designed tests to measure the efficiency of each dynamo and to see whose bulbs lasted the longest. In a room guarded around the clock by a security officer, bulbs from each of four rival companies faced off.

Today we think of the light bulb as one of the most standardized and interchangeable of household objects, but in 1884 and for many years after no two bulbs were quite the same. Skilled artisans blew each bulb by hand while others performed the delicate task of fastening the fragile filaments to platinum clamps inserted through each bulb's plaster base. To avoid patent infringement, companies used different processes and designs, but in each case the carbon filaments burned unevenly, weak spots developed and smoked the interior glass, and the lamp's power diminished the longer it burned; efficiency usually dropped 20 percent after only one hundred hours. The Franklin Institute conducted frequent measurements to see how well the competing bulbs performed over the course of days. As part of his sales pitch Edison had assured his customers that his lamps would last for six hundred hours, but after one thousand hours all but one of his twenty test bulbs still burned while more than half his rivals' had broken. Edison's bulbs proved less efficient than some others, but this did not prevent him from claiming victory. His competitors hotly contested the results, declaring that the test had been flawed from the start.[6]

. . .

The rival electric companies exhibiting their wares in Philadelphia were engaged in a ferocious struggle for patent rights and market share, but many who visited the fair could afford to take a more philosophical view. The exhibits seemed to confirm a common opinion, shared on both sides of the Atlantic, that their nineteenth century would be remembered as a turning point in human history, the moment when technological creativity became *perpetual*, not the occasional product of rare human genius but a fact of everyday life. New inventions seemed to be fast becoming the single most important driver of human progress, their power increasing exponentially each year as humanity scored ever more impressive victories over "the brute forces of nature." And these were triumphs shared by all, as each of these new machines and processes added to the "welfare, comfort and broader information of the people." Writers never tired of cataloging the most important inventions developed in their own lifetimes. For one, this included "sewing machines and mowing machines, steam engines, celluloid and dynamite, rock drills and rotary printing presses, vulcanized India-rubber and Bessemer steel, boring for oil and burning water-gas, photography and the phonograph, electric light and electric smelting, the telegraph and the telephone, the spectroscope and hundreds of other discoveries and inventions." Surveying all these things that had been unknown and unimagined by their grandparents' generation, nineteenth-century Americans could only wonder, "What can come next?"[7]

The honor for these victories belonged not to generals and statesmen but to scientists and inventors, whom one journalist called "the worthiest of heroes and the essential rulers" of nineteenth-century civilization. And the greatest figure in this new aristocracy of practical intellect was Thomas Edison, whose life story read like a chapter in Gilded Age America's favorite story: the rise of the self-made man. "The poor boy struggles up," as the story so often went, "from the chilliest sort of penury, to opulence and power, and displaces those of inferior skill in finance and trade." While many admired the "luck and pluck" of some of America's most powerful industrialists, financiers, and robber barons, these men often inspired as much dread as admiration. But the biographies of inventors offered a less ambivalent morality tale, a thrilling narrative about the "busy brain of the tireless inventor," or a fairy tale in which a dogged genius coaxes Nature to yield its "golden secrets."[8]

The inventors' story put the best possible human face on the emerging order of industrial capitalism, focusing on its enormous creativity, the reward it offered to talent, and its contributions to human comfort, instead of its more troubling disparities of wealth and power. Edison might well be "the self-made electric king of the nineteenth century," but he seemed a humble and benevolent monarch. Reporters found him not in some Newport mansion or Fifth Avenue palace but in the humble farm buildings at Menlo Park, where he toiled in greasy work clothes, played practical jokes, chomped cigars, and extolled the virtues of pie. None doubted his voracious ambition, but it seemed to be for invention, for the ingenious conquest of nature in service of humanity, not for personal aggrandizement and power. As Edison told one reporter, "My one ambition is to be able to work without regard to the expense. . . . I want none of the rich man's usual toys. I want no horses or yachts—I have no time for them." Many of Edison's rival inventors offered journalists a similar narrative; they were self-taught mechanics and tradesmen who succeeded by creating rather than by scheming. But none caught the public's imagination the way Edison did.[9]

At the Philadelphia exhibition, a sculptor tried to capture the growing Edison legend in bronze, creating a statue depicting what the artist described as "the moment" when the inventor had discov-

Those celebrating the close of the nineteenth century loved to catalog the many inventions transforming their lives. The light bulb protruding from the figure's forehead lights the way for another century of technological marvels.

....

ered the secret of the electric light. Conveying the sort of "eureka" experience that we now associate with the light bulb, the artist showed Edison connecting two wires, presumably about to send current to his first working bulb. This suggestion that Edison had discovered incandescent lighting, and in a flash of solitary genius, provoked scorn among other electricians. One journal sarcastically suggested that, while the artist was at it, he might as well make two more sculptures depicting Edison inventing both the telegraph and the lightning rod. Even Edison's closest laboratory associates shook their heads over what one called a "species of glorified mist" that surrounded the great inventor.[10]

Most electricians expressed at least a grudging admiration for all that Edison had accomplished. But they also knew that the scientific principles behind electric lighting had been worked out across decades in many countries; that the crucial components of the lighting system likewise emerged from the insight and toil of many; that Edison's own creations owed much to the talent and dedication of the team he had assembled at Menlo Park; and that several of his competitors had arrived at working systems almost simultaneously, in some cases holding key patents ahead of Edison. Or at least that was their claim, in a rat's nest of lawsuits that would take the courts many more years to resolve.[11]

Arguments over who would win the valuable patents for incandescent lighting mattered a great deal to the inventors, investors, and patent attorneys, but made little impression on the popular mind, which found the Edison legend irresistible and soon began to refer to all incandescent lighting as "Edison's invention." The public liked to think of inventors as solitary geniuses and folk heroes. But the people most invested in America's Industrial Revolution generally recognized that invention is a complex social process that can be encouraged or stifled by public policy. Accordingly, legislators, scholars, and educators investigated the springs of technological ingenuity, hoping to understand more about how it was done. Many believed that the United States offered the most important case study. Europeans often conceded that Americans displayed a remarkable aptitude for invention, particularly in the field of labor-saving devices. The country had not produced many philosophers, as one Englishman put it, "but her practical men may be numbered by hundreds. If a Yankee has an idea, he likes to put it into practice. He is not content to read a paper, and let someone else work out his theories." Pioneers in industrial invention, the British still made better products, and sold the world many more of them. But they conceded, with evident concern, that "it is from America that all the new inventions come to us."[12]

While the young nation was still far removed from the European capitals of higher learning, America had become a leader in technological innovation in a number of fields and the most enthusiastic marketplace for all new technologies, from powerful labor-saving industrial processes to time-saving household gadgets. As one educator put it, "very great inventiveness" had become a defining national trait, as many thousands of American workers devoted themselves to mastering "the untried possibilities of nature." How had this happened?[13]

Edison himself was not much help in answering this question. When reporters asked him why Americans seemed so much more inventive than their English counterparts, he proposed that the British did not eat enough pie. Only a bit more helpful was his frequent suggestion that inventive genius was nothing more than "hard work, stick-to-it-iveness, and common sense." However admirable those traits may be, Edison shared them with countless men and women who never filed for a single patent. Others who spent time with him hoping to uncover the secrets of his success noted his wide reading habits, his "vast patience in logical deduction," and an ability to make careful calculations while conducting "unlimited experiment."[14]

Many explained America's inventiveness as a by-product of its expanding market and its chronic labor shortage, which exposed so many workers to the era's new machines and encouraged the search for more technological shortcuts. Europeans had produced the first generation of many nineteenth-century technologies, but their producers enjoyed an abundance of cheap labor that undermined the economic incentive for greater investment in new machinery. "A man is driven to invent machines for saving labor," as one put it, "because he has so much trouble in getting labor, and it is so expensive when he does get it." Many found this creative drive to be particularly strong among those "inventive animals" who hailed from New England. "No matter what his training or what his calling," an Englishman observed, "his mind is working in a kind of backyard over some idea for economizing labor that is near to being realized." In the same way, some suggested, American housewives had created so many useful household tools because they lacked Europe's ready access to cheap domestic help, a problem that forced women, quite literally, to rely on their own devices.[15]

Others attributed American inventiveness to the nation's more democratic educational system, which offered many citizens the basics of a common school education. European universities might dominate the field of pure scientific research, but many of the era's most important inventions came from working people—farmers, artisans, and manufacturers who grappled with tangible problems in their everyday lives. Some lamented that America's common schools remained mired in a traditional curriculum of reading, recitation, and simple arithmetic, and thus did not provide the building blocks of a good mechanical education. But American workers had the literacy skills to take advantage of a growing popular literature on science and technology, and at least the rudiments of practical math. Perhaps more important than any academic skills, this broad education encouraged an independence of mind, a self-confidence that many considered a hallmark of America's "free and educated people." Unlike his European counterparts, as many put it, *the American mechanic is a thinker.* "He is not content to go on doing as generations of workmen have done before him. . . . He enters with spirit and enthusiasm into the mastery of details, he studies with earnest zeal the conditions of the problems before him, and he sets about the solution of the difficulty with thoughtful common sense and strong determination."[16]

In many cities, this democratization of knowledge was also nurtured by scientific societies, voluntary organizations that encouraged public interest in the latest breakthroughs in science and technology. These clubs built scientific libraries and hosted lectures on "interesting practical subjects," drawing crowds of hundreds for talks on everything from "electric propulsion" to "The Hopi Snake Dance." Some of these institutions developed more specialized "Electrical Clubs" that offered young men evening courses in the new science. A few of the larger ones included well-equipped laboratories where members could try their own experiments—all that an aspiring electrical inventor might need to try out a new idea.[17]

In an era increasingly eager to link cultural traits with heredity, others explained American ingenuity as an expression of the "race," a biological imperative that had made Americans "a great army of *Doers*." For those who pursued this line of thinking, white Americans traced their bloodlines back to the Anglo-Saxons and the Germanic tribes of northern Europe, a mythical heritage that was thought to account for their "irresistible thirst for action and achievement." Now that Americans were mopping up the conquest of an entire continent, invention seemed to offer them a next frontier, an outlet for their innate hunger for the "conquest of nature."[18]

While each of these explanations for American inventiveness had its advocates, many more credited the nation's democratized patent laws for its citizens' remarkable technological creativity. "A new power of achievement has come into human thinking," the head of the patent office observed, "and no explanation for the change is even plausible which ignores the stimulating influence of a century of patent law." Who could tell how many great ideas down through history had been left "buried in the napkin" or carefully guarded by craftsmen as guild secrets? But in the United States, unusually "liberal" patent laws gave intelligent artisans, mechanics, and farmers the protection that they needed to bring these into the open, temporarily enriching themselves while making a lasting contribution to the commonwealth.[19]

Recognizing the value of patent laws to national economic development, the drafters of the U.S. Constitution authorized Congress to "promote the Progress of Science and useful Arts, by securing for limited Times to Authors and Inventors the exclusive Right to their respective Writings and Discoveries." Even so, some of the most inventive of the founding fathers doubted the value of patents in a free society. Under the English system that influenced American thinking on this matter, Parliament gave the king the power to grant patents to "the true and first inventor" of a new product. While the king was supposed to use this privilege to encourage useful improvements either created or imported by Englishmen, he often used it instead to grant trade monopolies to his cronies and raise money for the royal treasury. Sensitive to this abuse, America's revolutionaries sought a more radical and democratic approach, some suggesting that the state had no business interfering in the world of ideas. Benjamin Franklin refused to patent any of his many useful inventions; a practical device that improved human happiness, he believed, should be given "freely and generously" to the whole human race. "As we enjoy great advantages from the inventions of others," he wrote in his *Autobiography,* "we should be glad of an opportunity to serve others by any invention of ours."[20]

Thomas Jefferson agreed, and never patented his own inventions either. But through a twist of fate Congress appointed him to lead the nation's first board of patent examiners, a job it delegated to the secretary of state. Jefferson carefully reviewed the nation's first patent applications, for a time storing the patent models under his bed, and soon came to see patents as a useful "spring to invention." Still wary of the system's potential abuse, he developed strict guidelines for the grant of a patent that had profound consequences for the development of American technology. Each application should be carefully reviewed by an inspector, he insisted, who should grant a patent only to the individual responsible for a new invention, and only for devices that were demonstrably new and useful. These founding principles were adopted in the 1836 revision of the patent law that created what one historian has called "the world's first modern patent institution."[21]

Under this system, the pursuit of a patent became an abiding preoccupation for many nineteenth-century American men and women. Because the American system granted patents to practical improvements in already established inventions, every farmer and mechanic who worked with a machine could see its limitations not just as a source of frustration but also as an opportunity to make improvements, and maybe even a fortune. The British government charged high fees for patent licenses, one part of a "convoluted" bureaucratic system that favored the wealthy and well connected, and "high-value, capital-intensive inventions." The American government charged just a tenth of that price for an application, and gave an equal hearing to applicants from the working class who might lack scientific training but knew how to "transform their thinking into things." For a "mere pittance," as one patent officer explained, even "the humblest citizen" could hope to turn a good idea into a temporary but valuable government-protected monopoly. As one historian puts it, under the American system "genius was redefined as the province of the many, not the rare gift of the few."[22]

The government's patent inspectors passed no judgment on an

The U.S. Patent Office, an engine of invention and a tourist attraction. A journalist declared it "a wonderful school of human nature. Here genius soars to its highest, and cupidity descends to its lowest level."

invention's potential success in the marketplace. The applicant only needed to show that he or she was the original inventor and that the device actually worked; until the 1870s, that meant providing detailed specifications of the device as well as a working model that was available for public inspection in the patent office. Just as the U.S. government nurtured domestic industries by erecting high tariff barriers, its patents granted inventors and their investors a temporary shield from the pressures of the free market while making the details of their insights a matter of public record for all to emulate.

To be successful, the inventor needed to use that limited period of monopoly efficiently, laying the groundwork for the development of a viable commercial product. Final judgment on an invention's value was left to the marketplace, which rewarded some fabulously, others modestly, and in the end dashed the hopes of thousands who staked their dreams on an idea that might just as well have been left "in the napkin."

While many observers praised the patent system as one of the great institutions of American democracy and an engine of invention and prosperity, others in the Gilded Age called for reform, or even the elimination, of patents. Some complained that patents raised technical and economic roadblocks for many inventors. In the process of creating a new device, an inventor often felt compelled to invent "around" previous patents, a detour that sometimes provoked great new insights but more often wasted time, producing redundant solutions to technical problems that had already been solved. Others argued that the nineteenth-century flood of patented inventions only encouraged the Industrial Revolution's tendency to subordinate men to machines, flooding the market with mass-produced products that were cheap but inferior while robbing artisans of their living. Like capitalism itself, the invention process was a form of creative destruction that kept the economy in constant turmoil. The electric light, to take just one example, threatened the livelihood of gas and kerosene producers and drove lamplighters to extinction, just as their technology had beached America's once great whaling fleet. Similar stories could be told about the turbulent impact of new inventions on most of the nation's important industries.[23]

Many farmers resented the patent system, which granted monopoly control to the inventors of such essential tools as barbed wire. Farmers were also vulnerable to lawsuits or obliged to pay royalties for almost every new device they bought to lighten their burden. In these years mail-order houses and traveling salesmen saturated the countryside with patented labor-saving devices, from corn huskers to premade coffins. But unhappy customers found that this system forced them to pay additional patent fees, a racket that could be ruthless and often fraudulent. Even when those collecting patent royalties were honest, consumers resented what they considered to be an unfair tax that lined the pockets of distant corporations. Patents seemed like one more abuse of power by an urban elite, and Granger and Populist platforms often called for their abolition.

Objections also came from those inventors who felt they had every ingredient for success except enough capital to bring their idea to market. A patent was useless unless one could sell it to a network of people willing to take a financial risk in exchange for a share of the profits—landlords and bankers, suppliers and sales agents. And once the invention had been launched into the marketplace, the inventor who did not sell his patent rights to a manufacturer had to use the courts to guard his or her intellectual property. Even successful inventors like Edison chafed at a system that swallowed enormous amounts of their time and money, forcing them to defend their claims in the patent office and in court. As one inventor summarized the problem, "What is the use of a patent if it proves to be merely a somewhat expensive admission ticket to a colossal litigation?"[24]

But despite these concerns, many Americans considered their patent system a prime cause of the nation's remarkable inventiveness. While many rags-to-riches stories in the Gilded Age affirmed the virtues of hard work and strong moral character, another version emphasized the life transformation that awaited the man or woman who made the better mousetrap. Patent laws create "opportunities of obtaining wealth and power that are not presented by any other set of circumstances under the sun," as one put it. "In a single moment of time by a quick operation of the mind, the poor mechanic may be put into possession of an idea that by its own development may place him among the princes in the land." Legend had it that the man who first put an eraser on the end of a pencil had earned enough from the idea to shod his horses with gold shoes. Here was a "lottery ticket" that gave every "wide awake" working person a chance to make better money with their brain than they ever could with their brawn. In this, the patent system served as the greatest American university, *Scientific American* thought, doing far more than any schoolteacher could to stimulate "mental vigor, intensity, patience and fertility."[25]

As American inventiveness exploded through the late nineteenth century, so did pressures on the patent office, which introduced reforms, including the addition of more and better-trained inspectors who were charged with handling the increasingly complex technical questions over originality that emerged in specialized fields like electricity. Even as Americans adapted their patent system to new challenges posed by evolving technologies, other countries acknowledged the success of America's patent system and revised their own to match. Eager to understand and master Western models of technological modernity, a Japanese delegation toured the United States in the 1880s, searching for the secret to its rising economic power. "We found it was patents," they concluded, "and we will have patents."[26]

The public debate over the merits of the patent system entwined with the largest and most important conflict of the day, the growing struggle between labor and capital across the industrial world. Socialists charged that while inventors deserved credit for creating the machines that fired the era's great Industrial Revolution, greedy corporations had soon learned to harness and exploit this creativity not to serve mankind but for the "enrichment of the few." They predicted that under a socialist system that banned the profit motive, workers would become even more inventive, since they would enjoy better public education and more leisure time for "the free and fruitful play of the mind." People invented to satisfy a natural creative urge, the socialists insisted, and out of a desire to help others. But capitalists bought up patent rights and used them to stifle competition by charging high licensing fees or threatening to sue their competitors into submission, effectively monopolizing knowledge in order to serve one corporation's private interests. In the socialist utopia depicted in Edward Bellamy's bestselling novel *Looking Backward, 2000–1887*, the nineteenth-century time traveler found that inventive geniuses in the better world of the year 2000 created new ideas simply for the chance to serve mankind, the reward of more time to pursue invention, and the dream of winning the society's highest reward: the honor of sporting a red ribbon that marked the wearer as a great human benefactor.[27]

Critics of socialism derided this view of the invention process as hopelessly naïve, reflecting an "astounding ignorance of the world of modern industry." Whatever its flaws, they countered, industrial capitalism could not be faulted for its failure to encourage inventors, nor for its inability to provide them with the resources they needed to bring their ideas to market. Perhaps a few eccentrics invented for the sheer love of it, they conceded, but for most, "their aim is the dollar." "Inventors in actual life," as another put it, "are generally distinguished by an insane desire for money [and] by the wildest overestimates of the wealth which their inventions will ultimately bring them." Remove the profit motive, they warned, and society would lose the benefit of many new inventions, grinding progress to a halt.[28]

And nowhere in the industrial world did this seem more obvious than in the United States, where political democracy and the economic free market seemed to liberate so many people's hunger for riches, fame, and power. Americans were fast passing the Old World in technological creativity, many argued, because the U.S. economy offered these rewards not just to a small educated elite and those who inherited titles of nobility, but to all those entrepreneurs who served their fellows through the marketplace. The patent office became a key mechanism for accomplishing that goal, a servant and gatekeeper for America's technologically talented that harnessed individual ambition and genius for the benefit of all. In a variation on the American dream, some people joked that every true American man would feel ashamed to go to his grave without at least one patent to his name.[29]

By the time of the Philadelphia exhibition of 1884, inspectors at the patent office found themselves deluged with applications from the inventors of electrical devices of all kinds, "new uses of the modern marvel." Before the late nineteenth century, the government had classed all electrical devices as "philosophical apparatus," curios and tools for the laboratory. But now hundreds of Americans each year stepped forward to offer new electrical inventions, hoping to earn profits and fame by making a contribution to this new industry.[30]

Some who followed the new claims by inventors complained that too many were drawn to the same small set of problems. A great "captain" of invention, such as Edison or Alexander Graham Bell, would provide the world with a conceptual breakthrough—then, for years afterward, an army of "privates" followed along this increasingly crowded path, offering incremental improvements in the new technology. Surveying the government's weekly bulletin of new patents, one noted that "bottle stoppers have occupied the thoughts of thousands, and the name of those who try to improve on the sewing machine is legion." But the American patent system's support of these more modest improvements in existing machines paid great benefits. As people used these groundbreaking new technologies they identified inefficiencies and shortcomings, and felt empowered to attempt improvements in the details. While many of these inventions failed to find a market, enough did to drive a cycle of perpetual development.

In the electrical field, inventors found ways to make dynamos and batteries ever more efficient and powerful, and suggested better forms of wire, insulation, and meters, more precise current regulators, and a wide range of new light fixtures and reflectors. The incandescent bulb became the locus of constant innovation. By 1893, Edison's bamboo filaments gave way to cheaper and more uniform loops made from an extruded cellulose paste, using a technique pioneered by the English inventor Joseph Swan. Manufacturers offered bulbs in a wide range of decorative shapes, from stars to candle flames to long strings of "peas"; bulbs that were frosted or silvered or tinted; and bulbs with multiple filaments that could be adjusted to give several levels of light. When light bulb theft became a widespread problem for hotels, another inventor patented a locking light socket. Manufacturers developed mass production techniques that made bulbs more uniform and reliable while steadily reducing their price, from a dollar in 1881 to seventeen cents just three decades later. The electric industry's growing demand for natural resources also stimulated rapid growth and invention in related fields, including a surge in the production of American copper, supported by new applications of electricity to the smelting process. While many praised Edison in 1882 for developing a lighting system that seemed perfect in every detail, what he actually launched into the world was the first draft of a perpetual work in progress that stimulated creative energy in a wide and expanding number of fields.[31]

While most American inventors devoted their energy to creating

these incremental improvements in existing technology, others still dreamed of making major breakthroughs. Every few months reports circulated across the country predicting that Americans would soon enjoy something new under the sun, the brainchild of a new Edison who was on the verge of transforming modern life while turning some lucky investors into tycoons. From the small Maine town of Skowhegan, for example, a local newspaper sent word about a young man who had concocted a new arc light "which he is confident will revolutionize electric lighting all over the world." The nature of this revolution was never spelled out but public demonstrations had impressed local investors. As the editor put it, "the invention is creating quite a stir where it is understood," and the backers felt sure they had hold of "the greatest light in the country." Like so many others, this man's "electrical scheme" disappeared from view as quickly as it had appeared.[32]

Industry insiders sometimes mocked the hopeful inventor who jealously guarded his secret machine, sure that it would soon make him a millionaire. As one editor brutally summed it up, "If he is an honest man, he is a fool." They particularly enjoyed scoffing at naïve inventors who seemed oblivious to the fundamental laws of thermodynamics, designing dynamos that promised "150% efficiency" or trying to patent "modified beer taps" that claimed to harness the boundless electrical power of the earth itself. But most observers kept an open mind. In those years, too many amazing devices had already appeared out of America's workshops; too many skeptics had to apologize later for scoffing at some young inventor's grand promises. Far from laughing at the eager inventors with their new contraptions, the public eagerly lined up for the chance to see the latest inventions for themselves, a first glimpse at the machines that were making their future.[33]

Seven

Looking at Inventions, Inventing New Ways of Looking

In the late nineteenth century, Americans read accounts of new inventions in almost every issue of their newspapers. Thinking of themselves as a sort of "people's university," the best papers and monthly journals offered their readers long profiles of inventors, digests of science news, and daily reports on the latest technological breakthroughs. Some listed brief descriptions of patents issued each month, though this became unwieldy as the number of new inventions grew exponentially in these decades. The expanding public audience for news about science and technology was marked in 1872 by the publication of *Popular Science Monthly,* a journal that provided nonspecialists with the latest scientific ideas from Europe and informed readers about everything from fossils to the causes of dyspepsia. "Whoever cares to know whither inquiry is tending," the editor explained in its inaugural issue, "what old ideas are perishing, what new ideas are rising into acceptance—briefly, whoever wishes to be intelligent as to contemporary movements in the world of thought—must give attention to the course of scientific inquiry."[1]

Those who wanted more could read the weekly *Scientific American*, a favorite of many aspiring inventors. Or by studying an article in the Franklin Institute's *Journal*, an adventurous young reader might learn enough to try to create his own steam engine or dynamo. Workers in the new fields of electricity and telegraphy pored over lively trade journals that included detailed discussions of the latest advances in science and technology. Even popular magazines like *Godey's Lady's Book* served their readers' hunger for news about "progressive events," informing them about everything from statistics on mining accidents to astronomers' latest speculations about the geology of Mars, and never failing to include details about the latest advances in electric light and power. As the editor of another journal put it, "Very circumscribed must be the mind, and decidedly limited the vision of him who can take no interest now in both the actual and possible verities of Electricity."[2]

As new uses of electricity became a staple of the daily news, many Americans flocked to technical and trade expositions. Some, like the Philadelphia exhibition of 1884, showcased the fruits of electrical investigation. Others cast the net more broadly, none more so than the annual fair sponsored by the American Institute of the City of New York for the Encouragement of Science and Invention. Each summer starting in 1829, the American Institute gathered under one roof what it called "the diversified triumphs of American Industrial art at their very best—a spectacle at once interesting, instructive, and grand." Its annual fairs aimed to "encourage a spirit of discovery and improvement," providing a forum where inventors could share their ideas, learn by exchange with men of science and industry, and have their creations brought to attention "over the whole nation."[3]

At these exhibitions many Americans took their first close look at the inventions that were creating the modern world—steam engines and lathes, the telegraph and telephone, power looms and printing presses. As historian Alan Trachtenberg has put it, "The fairs were pedagogies, teaching the prominence of machines as instruments of a distinctively American progress." The New York fair's advertisement for 1878 captures some of the excitement that could be generated by industrial machinery: "Hoisting & Air Compressing Machinery! Great Steam and Vacuum Pumps! The Wonderful (speaking) TELEPHONE! Scores of Interesting Operations." Each year throngs of gaily dressed men and women crammed the exhibition aisles, seeking new sensations in this cacophony of material objects—the handmade and the machine-made, as well as the machines that would soon replace handmade objects of many kinds. They toured exhibits of fancy shellwork and wood carving, along with displays of false teeth, adjustable piano stools, knitting machines, and steam-powered presses. One year they watched a powerful steam drill dig an artesian well right through the floor of the exhibition hall, and held their ears at a demonstration of the latest foghorns. They saw new designs for ventilated garters, egg beaters, clothes wringers, and raisin seeders—a showcase of all the mechanical ingenuity that Americans were learning to apply to every aspect of their daily lives in their restless search for the tools to make work easier and the rest of life more pleasurable.[4]

In the early 1880s, electric light companies offered the fair's most popular attraction. Crowds formed to watch crackling sparks leap from the generator, and marveled at electricity's power to turn the exhibition hall into a "fairy land" of light. Long before the electrical revolution, lighting had absorbed the energy of many inventors. The fair showcased dozens of patented improvements in the oil lamp, and inventors had even made significant improvements on the age-old candle, developing the braided wick and cleaner, cheaper fuels such as paraffin. Among the first electric lights on exhibit at the New York fair was an "electric torch," singled out for distinction by judges who found this ancestor of the modern flashlight to be "very ingenious."[5]

While the government used patents to encourage invention, these private institutes adopted another strategy—the lure of monetary prizes, or "premiums," for exhibits that offered "articles of ingenuity and invention." Some derided this "wholesale showering of medals," but fair organizers saw this as an effective way to stimulate individual initiative and creativity, a process that in the end would provide "immense national benefits." Judges scrutinized each of the thousands of new objects on display each year, queried the inventor, and rendered an educated guess about its originality and its likely fate in the marketplace. They rejected some entries with scorn. Others they declared only adequate, "fairly representative of the state of the art," perhaps claiming "advantages claimed by many others." Those items judged "novel and exceedingly interesting" went home with a special commendation from the American Institute, a modest cash prize, and sometimes the interest of a potential investor or manufacturer.[6]

If these fairs were "pedagogies," then the range of lessons varied a great deal. Scientists, engineers, and technicians visited these expositions, and government officials used them as opportunities to review and summarize the state of the electrical art. Others drew from these expositions a patriotic moral; the glorious display of America's growing technological prowess and material prosperity seemed to confirm the virtue of its democratic institutions. As the mayor of New York declared in opening the 1883 fair, "Where free thought, free speech, free action and a free press are, where there are no learned or wealthy classes separated by impassable barriers from those who toil, *there* will ever be found the most varied and the most useful inventions and discoveries of man." Invention, it seemed, was a democratic art. The institute also hoped the fairs would inspire working men and women to try their own hand at innovation, emulating what they had seen when they returned to their homes, farms, and shops. As *Scientific American* explained, fair visitors should leave "more heartily in sympathy with the spirit of modern scientific investigation."[7]

But many who came to these fairs sought no lessons in science or politics or invention. They simply wanted to be amazed by something new. In the Gilded Age, rapid technological change disoriented people and disrupted their lives, but this experience of "bewilderment" was not always unpleasant; the fairs' promoters understood that the sensation could sell tickets to a growing audience that found new gadgets to be exciting and fun. As one visitor to the fair put it, these lavish displays of new technology appealed to "the thinking, as well as the unthinking, people of the civilized world."[8]

While the number of patented inventions grew each year, the public hunger for technological novelty soon proved a difficult craving to satisfy. Visitors complained if exhibits offered too much of the same old thing. Even the powerful dynamos, once so mysterious and exciting, held no more appeal by 1887 than "a good display of pumpkins at a country fair." Organizers knew that in order to keep the crowds interested, they needed to offer ever more novel and spectacular displays. By the end of the decade, when the American Institute created an entire hall devoted to the latest improvements in electric light and power, visitors seemed to be losing interest in these technical exhibitions—the wonder of the new had been replaced by the opacity of the complex. "Grand and interesting as this exhibition was to the student and to the scientist," the director lamented after the fair closed, "we regret to state that it was not of any appreciable interest whatever to the general public." Most flocked instead around the barker promoting a patent medicine that "cheers but does not inebriate." Or they enjoyed seeing "Marsden the Masher" and his new rock-crushing machine, or the "Anatomical Temple," a five-foot-high model of a Gothic cathedral that a chef had made out of the bones of chickens, geese, and other fowl. "The crying lack of the show is action," as one reporter complained. "People wish to see something doing." The ticket-buying public wanted the wonders of science and technology, not its mundane realities—the flash, but not the enlightenment.[9]

By the end of the century, the very success of America's technological revolution, its spread and saturation into so many corners of daily life, spelled the gradual decline of the grand scientific and mechanical fairs. Even as new inventions revolutionized daily life, the sponsors of these exhibitions struggled to weed out the peddlers of common patented devices and to attract truly novel inventions that might fill visitors with the desired sensation, the push and pull of informed bafflement. Instead, visitors felt that the annual fairs too often showcased the mundane pleasures of commercial humbug, not so different from what one could find along any boulevard in the city, the pages of popular magazines, or even from barkers at a country fair. The American Institute's annual New York fair, the largest of its kind, sputtered along through the 1890s, but visitors complained that "scarcely a thing is to be found in the entire exhibit that is not familiar or easily within the knowledge of all well-informed people."[10]

Reflecting on America's remarkable technological creativity, one science educator claimed that an invention's "greatness" was best measured by its power to generate ever more inventions. By this standard the electric light was one of the greatest. Some of the most important electric light innovations did not lend themselves to display at an exposition and were often ineligible for a patent. They involved new uses of the light, a tool for amplifying vision that served such a wide range of fields that no inventor could have anticipated them all. Even Edison could not have guessed at the many uses for the new light devised by police officials, urban planners, factory owners, and entertainment entrepreneurs. And specialists in a wide range of other fields adapted the technology for their own purposes, joining a broad creative process that might well be called "the social invention of the light bulb."[11]

Photographers, for example, almost immediately recognized the new light's potential to improve their art and their trade. Before the arrival of electric light, professional photographers relied on sunlight to expose their plates. They placed their studios on the top floor of buildings, under skylights, but remained at the whim of the weather—rain, cloudy days, and even the chronic urban problem of smoke and soot put them out of business many days each year. Photographers soon found that by bouncing electric light off reflectors, they could create a substitute for sunshine at any time of the day or night. Tradesmen and hobbyists experimented with lights and reflectors in countless variations, and shared their successes in hobby clubs and trade journals.

The light also figured in one of the abiding fascinations shared by nineteenth-century Europeans and Americans—their desire to see, and thus to know and conquer, any remaining blank space in humanity's mental map of the globe. Professors carried electric light in their hot-air balloons in order to observe the clouds at night. Spelunkers used the light to map underground caverns, soon followed by entrepreneurs who installed powerful lighting systems in places such as Luray and Mammoth, drawing thousands of tourists each year to visit these "show caves." Others poured enormous energy into penetrating the dark mystery of the ocean floor. As one historian of oceanography has put it, "It is difficult to express adequately human ignorance of the ocean's depths up to the mid-nineteenth century." Among the earliest attempts to apply the light to deepwa-

Photographers embraced electric light and experimented with ways to diffuse it on their subjects.

ter exploration was an underwater observatory constructed in Nice, France, that brought eight passengers more than one hundred feet below the surface. Little sunlight penetrated to this depth, but visitors peered out through a thick glass window to an ocean floor illuminated by a powerful electric beam.[12]

At greater depths, most information could be gathered only blindly. British and American expeditions mapped the ocean floor with plumb lines while naturalists dredged the bottom, where they were surprised to find thousands of new species, strange creatures living far from the remotest ray of sunshine and in extremely high water pressure. While dredging pulled up slow-moving bottom

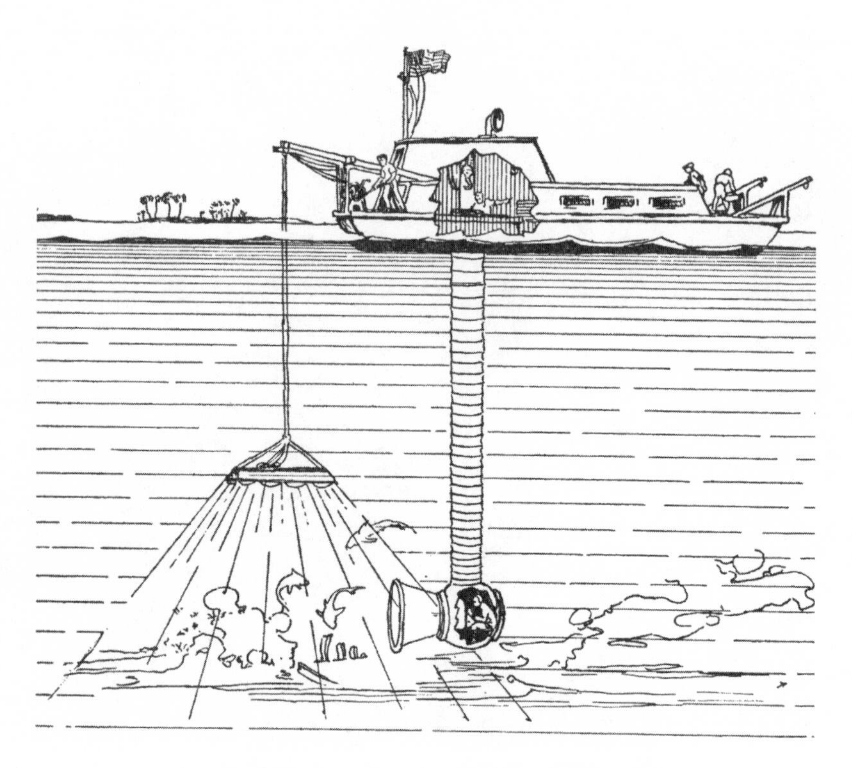

A 1915 experiment in underwater photography.

dwellers, Prince Albert of Monaco hoped to find something more. An avid amateur in the new science of oceanography, the prince led groundbreaking studies of ocean currents and gathered a world-famous collection of sea creatures. To reel in some of the deep's most elusive specimens, he rigged a battery-powered incandescent bulb as bait. Specially designed to handle both corrosive salt water and intense pressure, this "electric fish trap" proved quite successful in hauling up what one journalist called "rare specimens of the finny tribe in the superior interests of science," vastly expanding human knowledge of the deep sea.[13]

In those same years, the public paid rapt attention to another attempt to see, and so to conquer—a competition between various explorers determined to be the first at the North Pole. Edison followed the race himself, and offered the useless suggestion that explorers might try crossing the rugged Arctic ice fields by poling themselves along in five-hundred-foot-long, eight-ton iron sledges. While explorers had "no very sanguine hopes" about Edison's plan, they did welcome electric light as an important tool in their quest, and early expeditions used it navigate through Arctic fog and ice. The Norwegian explorer Fridtjof Nansen brought a windmill-powered cabin light on his 1893 expedition, using it to prevent the Arctic night's "awful monotony" from crushing the team's morale. Previous expeditions had tried to stave off boredom and despair by organizing "theatrical displays and musical entertainments," but this never worked as well as Edison's technical fix. "Electric lights are a grand institution," Nansen noted in his expedition journal. "What a strong influence light has on one's spirits!" In the end, however, he was forced to leave the comfort of his well-lit ship and travel by dogsled to within 230 miles of the pole, the closest that any explorer in the nineteenth century would come.[14]

Doctors immediately recognized the potential value that the new light might provide to the healing arts, another field profoundly improved by its ability to illuminate a once invisible world. The electric light amplified the power and consistency of the microscope, which in turn helped to confirm the germ theory of disease. Conventional microscopes had long revealed miniscule creatures in water and organic matter. The British scientist John Tyndall used a beam of electric light to demonstrate that the air itself carried an organic swarm of spores, bacteria, and other "excessively minute solid particles," a startling discovery that unsettled many, but added support to Louis Pasteur's germ theory. Electric light's power to reveal this hidden world proved valuable to health officials as they tried to publicize these radical new ideas about the source and prevention of devastating urban scourges such as cholera and typhus. At exhibitions and in lecture halls, scientists amazed their audiences by casting projections from their microscopes onto a screen using lantern slides illuminated by arc light. Even as nineteenth-century science asked the public to accept its claims about a world beyond human senses, the new technology gave some of this a tangible reality. "A drop of water presents the most extraordinary monsters imagination can conceive," one reported after seeing one of these germ slide shows. "Serpents, crocodiles, and worse dragons than St. George had to deal with, whirl about through their liquid element, striking terror to the hearts of all beholders."[15]

Surgeons embraced electric light technology almost immediately, another part of the late nineteenth century's revolutionary improvements in medical practice. "The nearer to the light we bring a diseased part," as they said, "the easier the diagnosis." Over the centuries doctors had tried using candles and mirrors to reflect light into the body's darkest corners, and performed operations under skylights. Experts recommended using the "cold north light" whenever possible since it cast less heat on the surgeon and fewer shadows on the patient. Doctors had also rigged devices to concentrate the beam of a candle or oil light, using it to illuminate translucent flesh and reveal the shadows cast by tumors, inflammations, and splinters—a blunt form of diagnosis, but in the era before antiseptic practices much safer than exploratory surgery.

In the earliest attempts to use electric light to improve surgery, some European doctors tried using a primitive lamp—a red-hot glowing wire connected to a battery—but this approach held obvious dangers for both the doctor and the patient. Incandescent bulbs were cooler, flexible, and much brighter, and doctors used them almost immediately to provide the first clear look at the living tissues of "the throat, nasal passages, bladder, and other portions of the inner man." Within a few years, instrument makers had crafted a series of specialized surgical lights, each adapted for the unique challenges posed by various surgical procedures. Doctors improved their powers of diagnosis, sending the focused light of incandescent bulbs into every opening in the body. For stomach ailments, some asked patients to swallow a tiny, pea-size bulb, then turned the current on, "reading" how the light passed through the body's "enveloping membranes." Dentists benefited as well, finding the light strong and cool enough to use right against the teeth and gums, illuminating defects otherwise hidden deep beneath the surface.[16]

At the close of the nineteenth century, many of these diagnostic methods were eclipsed by the German scientist Wilhelm Röntgen's discovery of X-rays, a still more powerful and mysterious form of "invisible light." The public reacted nervously to news of a light capable of peeking beneath the clothing, but medical experts embraced this discovery that, for the first time, provided a surgeon with "a map of the unknown country he is to explore." Along with the development of anesthetics and antiseptic practices, this new light laid the foundation for the remarkable accomplishments of modern surgery.[17]

Others explored the idea that the light itself might be good medicine. Not long before the electric light appeared, some health faddists on both sides of the Atlantic felt sure that blue light had unique healing properties. Those suffering from the late-nineteenth-century epidemic of nervous diseases and depression were advised to replace their clear glass windowpanes with blue glass, and a southern dentist swore by the use of an electric blue light for "the painless extraction of teeth."

The American health reformer John Harvey Kellogg scoffed at this "blue light fanaticism," but he pioneered the medical use of plain white electric light. Best known today as the inventor of corn-

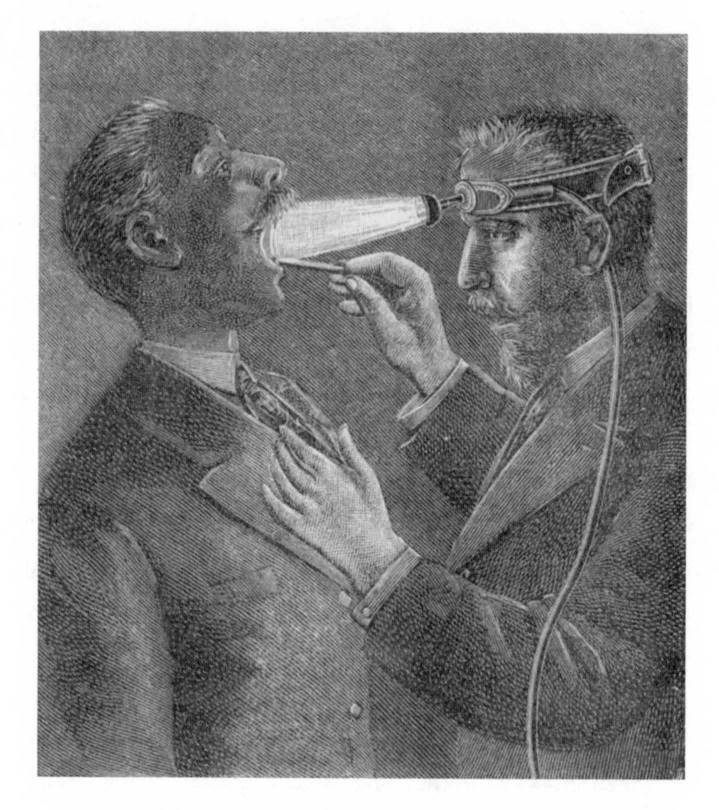

A late-nineteenth-century laryngoscope.

flake cereal, Kellogg ran a Seventh-day Adventist health spa in Battle Creek, Michigan, that catered to wealthy patrons, including J. C. Penney, John D. Rockefeller, and President William Howard Taft. While his brother William went on to make a fortune selling their cereal, John Harvey devoted himself to the healing arts, writing popular health guides that warned of the dangers of masturbation, and encouraging readers to adopt a strict regimen of simple food, fresh water, and plenty of exercise. In his "laboratory of hygiene," Kellogg experimented with the tonic effects of electric light. If the new urban environment disrupted sleep patterns and frayed nerves, producing a generation of Americans who retreated to sanitariums looking for a cure, then it was fortuitous for him to discover that the electric light, which had done so much to create these modern maladies, could also be used to cure them.[18]

Medical authorities had long recognized the healing properties of sunlight, and the German scientist William Siemens had published widely respected studies claiming that electric light stimulated the growth of plants. Following this line of thinking, Kellogg concluded that electric light was "nothing more nor less than a form of resuscitated sunshine." Thinking more poetically than scientifically, he explained that coal was a compressed form of the sunlight harvested by eons of primeval plants, an energy that was liberated when coal burned in a steam boiler; the steam turned the dynamo, channeling the electricity that created new light, "thus completing the cycle of utility." To sit under electric light, then, was to bask in the healing beams of untold millions of ancient days.[19]

Kellogg delivered light to thousands of patients, using what he called an "electric light bath." The bather sat in a small cabinet, its interior lined with mirrors and studded with sixty incandescent bulbs. In this way patients dipped themselves into a healing "sea of light." "Shed upon the nude surface of the body," Kellogg claimed, "the rays will enliven the nerves with renewed force, and will dissipate and destroy the innumerable malefic influences which imperil health and life."[20]

Kellogg conceded that much of the benefit of his incandescent bath was produced by the bulbs' radiant heat, not by the light itself. Like the traditional steam and heat baths already popular in spas, the electric light cabinet proved a good place to work up a sweat, stimulating the skin, accelerating respiration, and somehow encouraging the internal organs in their "eliminative work." Saturating the body in the warm glow of incandescent light, Kellogg claimed, would prevent disease, heal skin conditions, "nourish"

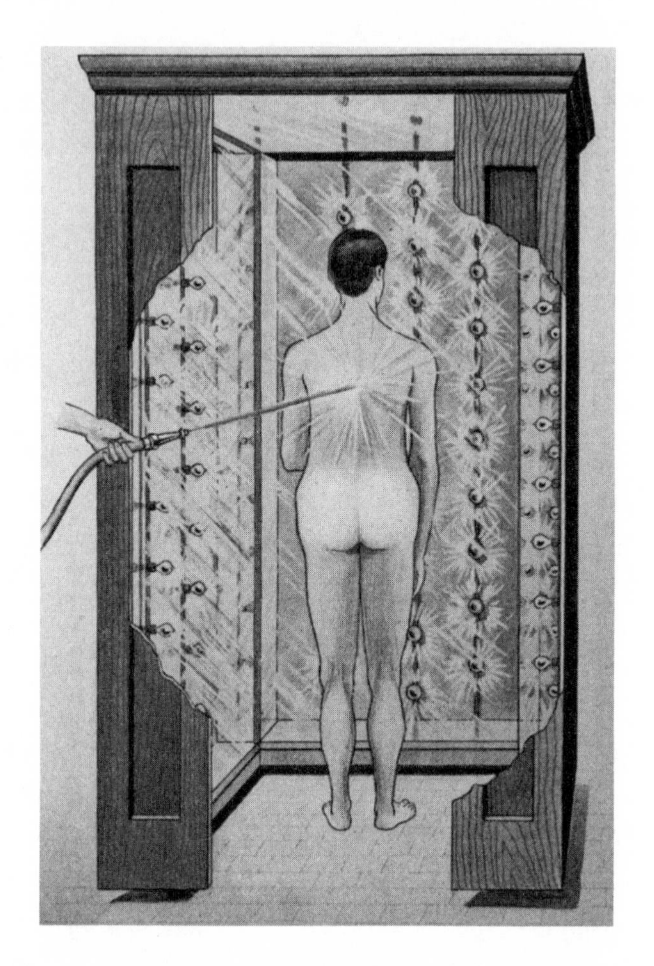

For some ailments, J. H. Kellogg recommended both an electric light bath and a "cold water douche."

the body, and was "the most agreeable means of reducing flesh"—especially when the bather followed the electric bath with a brisk rub-down with salt or ice-cold mittens. Other doctors followed Kellogg's lead, some claiming that they had successfully used electric light to treat syphilis, pneumonia, lead poisoning, diabetes, rheumatism, and "morbid conditions of the blood."[21]

The Russians and Turks had their own baths—Kellogg patriotically called his invention the "*American* bath," though it proved more popular in European hospitals and sanitariums; no well-appointed home among the European aristocracy was said to be complete without an "electric light cabinet" tucked in a bathroom corner, and the device enjoyed success at exclusive spas where those suffering from a range of maladies put their hope in this very modern form of relief.[22]

Kellogg's "light therapeutics" joined the many other "nature cures" embraced by suffering patients at the turn of the twentieth century. His sales pitch for the healing powers of electric light perfectly straddled two seemingly contradictory but deeply held ideas about medicine. On one hand, Gilded Age Americans felt sure they were making themselves sick by living in an artificial, man-made environment, and many looked for cures in a good dose of something "natural" like light. But many put equal faith in the latest remedies offered by science, hoping to find health in technology's power to manipulate the body in unprecedented ways. Kellogg's success as a peddler of the incandescent bulb's medicinal power reflects the appeal of a therapy that promised the health benefits of both nature and the very latest machine.

But in the 1890s many more Americans were struck less by the electric light's ability to heal than by its enormous power to destroy. The wealthy might retreat to Kellogg's spa to soak in the healing rays of incandescent bulbs, but in the streets of every American city the new technology unleashed a plague of fire and death by electrocution that left many wondering if the price of progress had grown too steep.

Eight

Inventing a Profession

In 1887, hundreds of electricians converged on Boston for the fourth annual convention of the National Electric Light Association, a trade association for the operators of the electric companies that had sprung up in every city and many towns in the previous decade. Friends of the fledgling industry had been urging this sort of cooperative exchange for some time. In the early years the business had been dominated by "hard-headed businessmen" eager to control any information that might give them an edge over rivals, inventors who jealously guarded their trade secrets, and electrical workers skeptical of outside interference with their business. These pioneers were "strangers to one another," as one editor put it, "working independently, with no attempt at harmony, with but little knowledge of one another's methods of business, with no established precedent to guide them." Since their business lacked any form of licensing or formal training, the first generation of electricians had learned their trade on the job, and many had the burn scars to prove it.[1]

But utility owners agreed that the time had come for better cooperation and communication, in order to ease some of the industry's growing pains. Electric light and power was fast becoming big business, with millions of dollars invested and hopes for many more. While gas still enjoyed the lion's share of the market, by the end of electric light's first decade more than 250,000 arc lights and three million incandescent lamps had been sold. Most large buildings going up now included electric wiring along with gas pipes, and it seemed just a matter of time before incandescent lighting would drive gaslight to extinction. Those who worked in the electrical industry might congratulate themselves for being part of "the most extraordinary development of this century," but many conceded that the industry's great energy too often verged on chaos. Rival electricians needed to work together in order to create common standards of safety and measurement, to impose some rational order on the ruinous competition of the marketplace, and to lobby government to advance their common interests.[2]

From the start, the electrical trade journals had encouraged this process of professionalization, helping their readers to think of themselves as part of a common cause. A half dozen of them, serving different regions of the country, reported on the latest research, arbitrated disputes among rival inventors and companies, advocated for safety measures and training programs, and shared information about contracts made and lights installed. Fierce partisans in the war on gas, they published any news they could find about gas explosions and asphyxiations and trumpeted milestones in the growth of electric power, such as the near-simultaneous decision by Queen Victoria and the emperor of China to have the new lights installed in their palaces.[3]

So the journals applauded when, in 1884, the leaders of some of the larger utility companies decided to organize the National Electric Light Association (NELA). As the *Electrical World* put it, for too long electricians knew each other "only in rivalry and dispute. . . . They never meet except by chance, and seldom if ever part as friends." NELA devoted itself to turning this group of suspicious individualists into professional comrades, hosting annual conventions that encouraged "loose but interesting discussions" and issuing formal reports on every aspect of the industry—subjects like the latest scientific discoveries from Europe, patent law reform, the relative merits of competing dynamo designs, and new ideas for creating demand for their product.

The technical discussions at these early conventions suggest the wide range of practical knowledge required by the managers of electric companies to keep even a small power plant functioning effectively. Both the coal-fired steam engine and the dynamo required regular and skilled maintenance. The dynamo's bearings needed oiling, and the massive leather belt careful adjustment; lack of attention to either would cause the machine to run inefficiently, wasting the company's money and risking damage to the dynamo itself. Work around the dynamo had to be done carefully since it revolved six hundred times per minute and if improperly handled could deliver a deadly charge. The operator also had to watch for any glitches that might cause fluctuations in the current. Too little current would annoy customers by producing dim or wavering light, but too much was worse, since it could burn out bulbs all across the line or melt safety fuses, shutting down the entire system—a common occurrence that led most electric customers to always keep their oil lamps handy. The electrician also needed to know how to install and maintain the grid of wires that branched from the heavy copper mains in the power station along miles of poles or conduits, and then into streetlamps, homes, and businesses through hundreds, and soon thousands, of fuses, switches, and lamps, organized in a growing variety of configurations. Finally, utility managers had to decide when to change each light bulb, since in those early days these were provided by the company, not the consumer. Instead of breaking, most bulbs just grew dim over time. Letting these feeble lamps burn might save the company money in the short term, but at the expense of undermining the customer's enthusiasm for electric light.[4]

Every one of these technical decisions had to be made with an eye toward costs. After all, as one conventioneer put it, electric company owners got into the business "not for glory" but for profit, and could compete only if they offered light at a price not much more expensive than gas. Most reported that business was growing, but this expansion introduced more technical challenges as companies needed to generate ever more power and distribute it efficiently over a larger territory. Utility owners had to learn all of this on the job, and soon realized how difficult it was to create and run a power network that kept pace with a rapidly changing state of the art. The inventors and manufacturers of electrical equipment were doing just fine, many observed, but those who carried the burden of running the daily operations of a light plant often found profits elusive. And so most came to the annual conventions with one question on their mind: "What are the best means and methods in order to take an electric light plant, run it successfully, make money and declare dividends?"[5]

Convention organizers knew just how to turn a bunch of suspicious rivals and competing electrical authorities into an "electrical fraternity," making sure that delegates had plenty of time for feasting and carousing. Along with some archery and target shooting, the electric men played baseball, a two-inning affair in which the lean men routed the fat ones. Like any other band of conventioneers, the group also found time for what one called "surreptitious games." As it did for so many other professions, this chance to cavort in a strange city provided a bonding experience that encouraged members to think of themselves as colleagues in a common cause: the electrification of America.

In the exhibit hall, NELA members visited displays set up by the

Manufacturers showcased their electrical wares at the annual NELA convention.

....

country's many new electrical manufacturing firms. Salesmen touted the particular virtues of their wire or lamp fixtures, while greasing the wheels of commerce by passing out complimentary cigars and souvenir badges. Back on the convention floor, the visitors enjoyed plenty of preaching of the new electrical gospel. To raucous applause, speakers heaped contempt on their gas-dealing enemies and painted an inspiring vision of an electrical future. "I say give us the opportunity of attaining the highest civilization we can enjoy," one industry leader thundered. "Give us sound and healthy bodies; give us no more darkness; but give us light! More light! Give us the electric light. It is the poor as well as the rich man's light. It will light the suburbs as well as the central portions of your city. It is, in fact, the light for all."[6]

When three hundred electric light men gathered for their 1890 convention in Kansas City, however, their faith in their industry's future had been shaken by a recent series of electrical fires and accidents that had provoked a public backlash. As usual, the convention began that year with a warm welcome from the host city's mayor, who slathered the electricians with the same sort of flattery that they had enjoyed so many times before. "What you and those of your profession have accomplished," he told them, "reflects credit upon your country, borders close upon the miraculous, and entitles you to imperishable honor and renown." The nation looked on with "wondering eyes and bated breath," he concluded, to see what miracle these electrical men would work next.[7]

But when the real business of the meeting began, industry leaders acknowledged that many in the public now felt more fear than respect, more anger than awe. Electricity had proven to be a powerful but "dangerous servant," and some wondered aloud if it was doing more harm than good. Newspapers admonished the electric companies for unleashing a destructive force that they seemed unable to control. The association's president considered this "present outcry" to be unfair; after all, many more had died in the early days of steam power and gaslight. Another speaker dared the government to investigate the industry's safety record, declaring that electricians had "nothing to dread but the misrepresentation of their foes." Still, the conventioneers agreed that they could not afford to ignore the "popular and largely sensational agitation" that was spreading across the country, inspiring some cities to chop down their wires and others to talk about government ownership of electric power. The delegates understood that their young industry faced a crisis of authority—if they did not impose order on the future of electric light and power, others would.[8]

Among the industry's recent setbacks was the embarrassing matter of having recently burned down a large section of downtown Boston. On Thanksgiving Day 1889, overheated wires at the Boston Time Electric Company spread flames through a building packed with flammable dry goods. By day's end, the fire had reduced much of the city's business district to ashes, killing two firemen and destroying millions of dollars' worth of property. Only a week before, the *Boston Herald* had declared the city's downtown buildings fireproof, but they had proven no match for electrical workmanship that the fire inspector denounced as "extremely negligent and shiftless." This was, he concluded, a "preventable disaster."[9]

The inspector insisted that the time had come for the government more actively to supervise the electric companies, ending the pioneer era of rapid and reckless expansion, a market so free that it had become "thoroughly anarchistic." Insurance companies, firefighters, and civic leaders all agreed, and the city gave its Board of Electrical Control more power to regulate the electric lines. City workers began to remove the vast collection of dead wires that dangled overhead and any live ones strung on poles and buildings that might pose a risk to firefighters. And they enforced new rules that required electric companies to upgrade their systems by installing the latest fuses. Any lighting that lacked these safety devices faced the city's "axe."[10]

Meanwhile, electric wires continued their mischief, starting several more fires around Boston and causing sensational accidents. A wire fell down across a pair of horses on Boylston Street, for example, knocking them both to the ground. "This will never do," the *Boston Globe* insisted, reasonably suggesting that the electric companies should be forced to hang their wires "so strongly that they cannot fall on people's heads." Public fear and rage spilled over into a stormy legislative session in which citizens demanded new state laws that would hold the lighting companies accountable for any destruction they caused, and tighten government control over the industry. Electric companies lobbied hard, and successfully, to defeat most of these proposals, but they knew that the controversy was far from over.[11]

The state's leading electric men did not resist all efforts to impose a more rational order on their business, agreeing to submit to more government inspection of their wires. But they were alarmed by the growing public fear and resentment of their services. When Boston's mayor granted additional franchises for street lighting not long after the blaze, citizens objected that they wanted "no more electric wires." They expressed particular resentment because the lighting contract had been granted to a New York company. "Electric construction in New York has been notoriously defective," they warned, "and has borne a shocking harvest of fatalities that are fresh in the public mind."[12]

The most shocking of these many New York fatalities had occurred just a month before the Boston fire. A telegraph lineman, John Feeks, was clearing dead wires from a pole when he touched one that had evidently crossed with a powerful arc-light wire. Though he died instantly, for over half an hour his body remained high above the street, trapped in the net of telegraph, telephone, and fire alarm wires. As comrades struggled to free his corpse, thousands of New Yorkers gazed up, watching flames shoot from the lineman's mouth and nostrils and roast his hands and feet. So many pressed in for a closer look that police had to drive back the crowd, which spilled off the sidewalk and blocked traffic. Meanwhile, nearby roofs and windows filled with people "impelled to gaze upon a sight that thrilled them with horror." The accident happened close to city hall, so many officials joined the crowd, including one who declared his intention to have all the electric wires in his district taken down at once.[13]

The next day, sympathetic New Yorkers lined up to drop donations in a tin cracker box that someone had nailed to the "fatal pole." Shoeshine boys chipped in dimes, workingmen their quarters, and even a blind beggar offered his two cents. Together they raised more than two thousand dollars to support the young woman now known across the city as "widow Feeks."

Even before this incident, the mayor had been waging skirmishes against the electric companies, trying without much success to enforce laws already on the books that required utility companies to bury their wires. Across the country, other city leaders had been engaged in a similar "battle of the wires." Chicago's city electrician declared war on all those arc-light companies that had thrown up poorly insulated wires "at random over the housetops," and by 1884 eleven miles of conduit had been laid under the streets, holding over two hundred miles of wire. When some electric industry leaders complained that these restrictions slowed the growth of their business, turning Chicago into a "dark city," the mayor replied, "Chicago wants electric lights—but not death." Cleveland's fire department, aided by a band of private citizens, removed poles and took down dangerous wires, declaring that "innocent citizens have some rights which even the electric light companies are bound to respect." The mayor of Philadelphia also waged a populist war against overhead wires, threatening to chop them down himself.[14]

In St. Louis, electric company officials and city leaders gath-

When Judge *depicted the death of John Feeks on its cover, a New York electrical expert complained that the picture slandered incandescent lighting: "This cartoon, for absolute disregard of fact, has certainly never been surpassed."*

....

ered on New Year's Eve 1889 for the ceremonial first lighting of a large section of the city, applauding when the lights came on without a hitch. Almost immediately, however, telephone customers complained that all they could hear on their lines was a powerful hum, like a hive of bees. Thanks to interference from poorly insulated electric light wires, the telephone network had become "full of electricity." Some switchboard operators received a nasty shock and the rest expected to "get it in the neck" at any moment. The stray current found its way into the police department's call boxes; when one patrolman answered his phone, it erupted in a sheet of blue flame and shot golden sparks until he doused it with a bucket of water. Investigators traced the problem to shoddy construction of the lighting system and short circuits caused by breaks in some of the hundreds of miles of rusting dead wires strung above the city streets. As a temporary solution, the electric company agreed to wait until 11:30 in the evening before turning on its streetlights so that the citizens of St. Louis could use their phones till then. Eager to protect their business and their customers, the telephone company pressured the city to force the electric wires underground, in spite of warnings from St. Louis electricians that such a move would be "the greatest mistake ever made."[15]

Though New York had passed a law requiring the electric light companies to put their wires underground, the companies responded by putting their "figurative thumbs to their figurative noses." Even if the technology existed for safely burying the wires, the electricians argued, too few underground conduits had been built. Desperate to keep their costs down so they could compete with the gas companies, the electric companies tried to pass to the taxpayers the cost of burying wires. If the city wanted buried wires, they argued, it should lead the way by tearing up the streets and building the conduits. Until then, the companies insisted that their wires belonged in the air where they could at least keep an eye on them. So many died from the overhead wires that the city coroner declared them a major health hazard, but the light companies refused to budge.[16]

Ignoring the electric companies' protests against buried wires, the city had begun digging up the streets and building conduits, but made slow progress. Electricians blamed this on "politics and jobbery," especially since the entire contract went to a single company owned by well-connected politicians. Some objected that tearing up the pavement was both a traffic nuisance and a health threat. The soil beneath America's urban streets was toxic, saturated with decades' worth of leaking gas and sewage. By one estimate, 10 percent of all gas sent through underground pipes seeped into the ground. Sure that the project would contribute to the spread of cholera, doctors warned that burying the wires would mean having to bury many citizens as well.[17]

The advocates of buried wires insisted that the technical challenges had been resolved, as did many entrepreneurs eager to get into the underground conduit business. But even on streets where the city did have conduits ready, the electric companies stalled and "through the ducts there passed nothing but wind." Some experts supported the companies' claim that burying wires offered no solution. While inventors had developed hundreds of patented ideas for protecting buried wires, a committee appointed to investigate the matter found that none worked well in preventing the ravages caused by ice and heat, leaching sewer gas and steam, and the shovel blade of the hapless ditchdigger. When some companies tried the experiment of burying arc wires, they found that current leaks reduced the efficiency of the dynamos and sometimes ignited pockets of leaking gas. Sidewalks erupted and heavy iron manhole covers shot through the air. At other times the current from failed wires boiled the water that leaked into underground pipes, sending geysers of steam up through the paving stones. Faulty underground wires also electrified streetcar rails, giving pedestrians and horses a terrific jolt. Residents of the Gilded Age city faced electrocution from above and explosions under their feet.[18]

. . .

Backed by a surge of public anger over the death of Feeks, the mayor of New York vowed to find the guilty company and charge its owners with manslaughter. The grand jury investigation, however, soon bogged down in confusion, further evidence of the chaotic nature of the city's wiring system. It took days to determine who owned the various wires on that pole, and in the end no one could tell who owned the pole itself. Two arc-light companies that were prime suspects denied any knowledge of a short circuit in their lines. They pointed their own accusing fingers at the city's Board of Electrical Inspectors, calling its members incompetent political hacks, dispensers of red tape that prevented the companies from conducting proper maintenance on their wires. In short, the electric companies blamed the death of lineman Feeks on too much regulation, not the lack of it.

Unable to find the culprit, the chief electrical inspector denounced the city's two largest arc-light companies, belatedly declaring them responsible for hanging five hundred miles of poorly insulated, "death-dealing" wire over the heads of New Yorkers. The mayor ordered city inspectors to check all wires and take down any not properly insulated, as well as those running along streets where underground conduits were available. In the dark week before Christmas, seven crews tore down miles of wire and hundreds of light poles. To ensure their safety, the companies even reluctantly shut down many of their properly insulated lines. In much of the city, residents found themselves thrust into a government-ordered blackout, their once bright streets now "caverns of gloom." Shopkeepers dusted off their old oil lamps, while gaslight flickered once more in Broadway theaters. By one estimate, a million and a half people now spent their evenings in "worse than Egyptian darkness." After two muggings on darkened streets, an editor declared the city to be "at the mercy of thugs." The mayor ordered additional police patrols and vowed to keep the lights out until the electric companies could assure the safety of their systems, a demand that the companies declared "the height of absurdity." "There were 35 deaths from gas in this city last year," the editor of a New York electrical trade journal complained, "but not a single newspaper has had space enough to mention that fact, their columns being crowded with grotesque details of one death by electricity." The Feeks incident was regrettable, electricians conceded, but "one would never think of asking that railroad cars should be stopped running because they now and then kill a careless crosser."[19]

The citizens of New York saw things differently, giving vent to a growing rage against the "murderous work" of the electric companies, "insolent corporations" that valued profits above everything else, even human lives. For years the industry fattened on profits while claiming "the privilege of murdering its linemen unpunished." Sending the manager of an electric company to jail for manslaughter, the *New York Tribune* concluded, would be "a present satisfaction and a permanent blessing to this community." The cover of one popular magazine captured the public's angry and anxious mood. *Judge* ran a cartoon depicting poor lineman Feeks trapped in a web of deadly wires, about to be devoured by the spidery "unrestrained demon" of electricity.[20]

At the height of this controversy, Edison himself weighed in. The nation's most respected authority on all things electrical, he only added a new layer to the controversy. He agreed with his fellow electricians that burying the wires offered no solution. In urban centers, Edison's system used heavy copper mains too large to hang on poles, so from the start he had buried many of his own wires at great trouble and expense. Still, he found that the technology for protecting them remained imperfect. Thus he argued that the arc-light companies' wires, which carried high-voltage alternating cur-

Cutting down poles and wires on Broadway.
The New York Times *reported that "after a long and tedious fight . . . New Yorkers were about to see one of their fondest dreams realized."*

Burying wires in New York City, 1889.

rent, could become even more lethal underground than they were overhead. Their insulation would inevitably break down, he predicted, and they would cross with otherwise harmless phone and telegraph wires, sending deadly jolts of electricity into homes, stores, and offices. "If a nitroglycerine factory were being operated in the city of New York," as he put it, "and the people desired to remove the danger, no one would suggest putting it underground."[21]

After that dire warning, Edison offered a self-serving solution to this problem, urging the city to ban all high-powered alternating-current (AC) wires, above or below ground, and to replace them with safer wires that carried low-voltage direct current. In short, Edison suggested that all companies should be made to use the system he had developed, one that relied on a force "so feeble that the wires, even at the point of the generator itself, may be grasped by the naked hand without the slightest effect." In this, Edison took aim not only at the arc-light companies but also at a new and serious rival in the light and power business, the central stations being developed by the Westinghouse and the Thomson-Houston companies. Their systems generated high-voltage alternating current and used transformers to step down the power for safe use by each customer. Already cutting into Edison's central station business by 1889, these AC systems were more efficient and flexible, and could serve customers across much larger areas. Edison had considered but rejected the use of AC, concluding that it posed insurmountable technical problems and could never be made safe. Citing "nearly one hundred deaths" across the country caused by arc-light wires, he declared this an "unanswerable argument" for banning all systems that used alternating current.[22]

Outraged by Edison's attempt to use the death of John Feeks as a way to damn their high-tension currents, George Westinghouse and his partners issued public statements to the city Board of Electrical Control, denouncing Edison's system as "dangerous in the extreme" and declaring his claims about AC power to be "grossly incorrect and even absolutely false." They challenged Edison to a technical showdown, a public contest on the relative merits of direct and alternating current, to be judged by a panel of disinterested experts. The controversy turned farcical when an Edison supporter challenged Westinghouse to a "duel by electricity"; he would take increasingly large jolts of Edison's direct current, while he dared Westinghouse to try the same with alternating current. The first to "cry enough" would have to "publicly admit his error."[23]

Few in the public could follow the heated, technical, and contradictory claims made by the rival companies and their paid experts, or the bickering between the city inspectors and the arc-light companies. The newspapers continued to print each round of accusations and joined citizens in muttering over this "electric light muddle." That winter the fear of electrocution tingled the spines of many New Yorkers, a palpable anxiety about "being slaughtered without notice by an invisible agent." As a journalist put it, "one scarcely ventures to put a latch key in his own door." The experience left some influential voices wondering if electricity could ever be properly controlled. "Its laws are apparently ill understood," warned the editor of *Harper's*, and the self-proclaimed experts who were selling it to the public seemed helpless to "control or subdue its antics." As competing electrical companies tore into each other and shirked responsibility for the calamities caused by their wires, friends of the industry worried about the public's loss of confidence in the new technology, and the sliding value of electric company stocks.[24]

The standoff between the city and the electric companies could not last long. While many newspapers cheered when the wires came down, they soon longed to have the old light back and blamed the mayor for his wire-cutting binge. "The work of destruction is evident enough," the *New York Times* wrote, "but is anything going on in the way of construction, or are we to sit in darkness and admit that street lighting by electricity is a failure?" The *New York Sun* likewise bemoaned the loss of electric light. People would choose a dim gaslit street over a brilliant electrified one that "exposes them to the peril of sudden death." But when the city's major thoroughfares returned to gaslight, everyone got a vivid reminder of electricity's "vast superiority." "We must have the electric light," the *Sun* concluded. "We cannot get along without it."[25]

Some members of the grand jury investigating the death of Feeks struggled to stay awake while listening to ten days of bewildering and contradictory testimony from city inspectors and the light companies. In the end they never found who was responsible for the accident and could only issue yet another plea for underground wires and stricter inspections. The state legislature held its own hearings on the dangers of electricity, where experts testified that the city's wires had been put up by "all sorts of irresponsible

A storm in January 1891 "demoralized electrical service" in New York and along the East Coast. To prevent further accidents, electric companies had to shut down service entirely, plunging city residents into darkness and isolation.

people; any man who had a little power to spare would put in a dynamo and run some wires, and those wires were not, frequently, insulated at all." Lawmakers proposed a variety of new regulations to deal with the perils of electric light.

Nature made its own argument for burying urban wires as a series of winter storms wreaked havoc on the electrical system, wiping out millions of dollars in capital investment in a matter of hours. Wires hastily tacked up by the electric companies came down in droves when coated with wet snow and ice. The poles soon followed, filling the streets with splintered lumber and "hopelessly tangled wires." To reduce the risk of fire after a severe January storm, New York shut down all of its "demoralized" electrical service, plunging the city into a darkness even deeper than the one produced by the mayor's war on the overhead wires. The only casualty of that storm was a horse killed by a live wire, but falling poles had smashed houses, lampposts, and trees, and it took many weeks to bring the lights back on. Facing immense losses, the electric companies conceded that here was "the strongest possible argument for putting the wires underground."[26]

And so when the electricians gathered in Kansas City in 1890 for their annual convention, their industry faced crises on a number of fronts—the clash between alternating- and direct-current advocates that would shape the future of the industry, the political pressure to put wires underground before the technology had been perfected, and the public's growing distrust of electricity and the companies that sold it. Each time a lineman died on the job or stray current caused a fire or a city plunged into darkness during a storm, the public's fear and resentment escalated—and sparked another round of calls for tighter regulation. The electrical journals protested against this meddling from those who scarcely understood the technical challenges and economic pressures the companies faced, and blamed the newspapers for sensationalizing the risks. But some industry leaders began to admit that their critics had a point; perhaps what many electricians dismissed as the "stupid fight against electricity" was not always led by ignorant Luddites. In their race to win franchises and string wires across the country, electric companies had ignored the public's legitimate concerns about safety and aesthetics. Eager to "crush their competitors," rival companies had hired unqualified men, used substandard materials, and bribed and manipulated city councils instead of winning the public's trust and support. One industry leader thought it was time for electricians to admit that their anarchic methods had "retarded the progress of electric lighting beyond our comprehension." Unless the electricians took steps to solve these problems and restore public confidence, they were only inviting "adverse criticism and adverse legislation."[27]

Making matters worse, many cities and towns expressed growing resentment about the cost of electric light. When the electric companies had arrived a decade earlier, civic leaders had eagerly granted them franchises in hopes of breaking the gas companies' monopoly. Many officials had naively expected the new light to be so abundant that it would soon become "practically free," ignoring the warning from skeptics who predicted that the tyranny of the gas company would only be replaced by the tyranny of an electric one. As one of these Cassandras put it in 1881, "There is not the least reason to believe that electricity exerts any better moral influence than gas, and we know that when a number of reasonably Christian men form themselves into a gas company they immediately become pirates of the most merciless and extortionate character. Why should we look for better things from the electric light companies?" A decade later, many agreed that electric companies were using their city franchises to steal exorbitant profits.[28]

By 1890, progressive-minded reformers urged local governments to solve the problems of electricity's safety and cost by going into the lighting business themselves, installing and running their own utilities. Pointing to many successful examples of efficient utilities run by city governments in Europe, they argued that private companies competing in a free market had failed to deliver essential city services at a reasonable cost, and that "municipal ownership" was a better way for the public to procure clean water, sewers, fire protection, trolley service—and gas and electric lighting.[29]

Some recent and spectacular government scandals made many Americans wary of handing over more tax dollars to city politicians. None could forget the Tweed Ring's plunder of New York City's coffers in the early 1870s, just the most notorious of many abuses of the public trust. Running a complex technological system like a power plant required technical expertise and business savvy, talents not often found in men drawn to the bare-knuckled world of urban machine politics. Critics of municipal ownership feared that letting city governments into the lighting business would only line the pockets of corrupt politicians and make cushy jobs for their unqualified friends, lazy "barnacles" who clung to the lumbering ship of state, "drawing pay without work."[30]

But in the late nineteenth century, a new generation of socialists and progressive economists offered a range of ideas for making city government a better steward of taxpayers' dollars. That involved taking many decisions out of the hands of elected politicians, entrusting them instead to technical experts who were presumably more devoted to science than self-interest. The "true source of municipal corruption," they argued, was wealthy businessmen who used their influence to win lucrative franchises by buying politicians, controlling public opinion through the newspapers, and other "unscrupulous devices." As one urban reformer put it, those private light companies that operated under the grant of a city franchise "are never willing to let well enough alone. They are constantly seeking new favors, and for this purpose require influence among government officials." By this view, municipal ownership of utilities was not an invitation to corruption, but its cure.[31]

The demand for municipal ownership of electric lighting and other utilities became a central goal of the "Nationalists," a reform movement inspired by Edward Bellamy's novel *Looking Backward, 2000–1887.* Hoping to realize Bellamy's vision of a future utopia built on a marriage of technology and socialism, thousands of Americans formed Nationalist clubs across the country in the early 1890s. Reacting against the social strife and wasteful practices that seemed endemic to industrial capitalism, they vowed to work toward a future in which all industries were "nationalized," converted into benign government-owned monopolies that served "the organic unity of the whole people." The electric light and the telephone were wonderful inventions, Bellamy wrote, "but what good have they done the people?" America's technological creativity had only served to concentrate wealth and power in fewer hands, forging an industrial tyranny that made living conditions worse for most working people. "The arc-light reveals scenes of squalor, misery and human degradation," as he put it, "which the tallow candles of our fathers never witnessed."[32]

While Bellamy's Nationalists wanted the entire economy to one day come under public control, they thought the federal government should start by taking ownership of the basic infrastructure of commerce—the railroads, banks, and coal mines. The movement enjoyed more success at the state and local level, where its advocates urged governments to start down the road to "nationalism" by owning their own utilities. Aided by the public anger against the electric companies in the aftermath of the Boston fire, Nationalists in Massachusetts won a bitter two-year battle in the legislature and the courts. The Municipal Lighting Act freed towns in that state to experiment with running their own gas and electric plants. When citizens came together to make their own light, Bellamy argued, they not only broke the stranglehold of selfish corporations but, as importantly, started "thinking along the line of municipal self-help."[33]

Bellamy's Nationalist movement faded within a few years, though its call for government ownership of key industries lived on in the platforms of the Populist and Socialist parties. At the same time, many Americans who had no interest in radical politics supported the municipal ownership of utilities. Towns often ventured into the field by building their own water plants, and by 1890 hundreds thought it made economic sense to expand into the new field of electricity. A technology that only years before had seemed a freakish novelty, and then a luxury enjoyed only by the most fortunate, was fast becoming something that city residents considered a public necessity, too essential to be left to the whims of profit-seeking corporations. Suspecting that the private companies were reaping enormous profits while taking shortcuts that endangered human life and property, citizens in hundreds of cities and towns created public lighting companies.[34]

Advocates of municipal ownership published introductions to the economic and technical challenges of running a utility, designed to give citizens the information they needed to make an informed decision about electrification. "The people have never had an inning," as one put it. Confused by the complexities of the new technology and the self-serving obfuscations of the electric companies, towns had been paying for their light "not on the basis of what they could buy it for, but on the basis of what the company could get." Reformers compiled statistics showing the wide range of prices that various towns paid private companies for their lights. Once public officials began to compare their annual light bills, many were shocked to find that they paid a great deal more than others, and they felt gouged.[35]

The lighting companies offered reasons for the wide variation in price. The earliest installations had become outdated and less efficient; the price of coal or water power varied considerably around the country; some towns wanted well-lit streets all night long, others only till midnight, while many frugal ones contracted for a "moon schedule," using no light during full-moon evenings. Given these and other variables, companies and customers were bound to disagree about how much light a town needed and how much it was worth. But many towns that tried municipal ownership reported that they enjoyed more light for less money, while eliminating the corrupting influence of the electric companies' "sharp, shrewd men" on their local politics.[36]

The municipal ownership movement received enthusiastic support from a new generation of academic economists, organized as the American Economics Association. These maverick social scientists challenged the orthodoxy of free-market capitalism, declaring that "competition is not always a good thing." The chaos in the field of electric power proved a case in point. In some towns, as many as two dozen rival companies fought for business. The result was not lower prices but shoddy and dangerous work, economic pressures that produced poor service at high prices, and the eventual absorption of small companies by their larger competitors. Even in cities where several companies split the business, they ended up charging the same price, evidence of price fixing. Economist Richard Ely called the electric light business a "natural monopoly," a field where the lowest price and best service could be provided not by many rivals competing in a free market, but by a single company large enough to take advantage of a greater economy of scale. Ely argued that city and town governments should not give away this valuable lighting monopoly to a private company but should keep it for themselves, saving tax dollars by hiring their own electrical experts to run municipal power stations.[37]

This sort of talk incensed the electricians who gathered in Kansas City for their convention in 1890. "Who are the great apostles of municipal ownership of electric lighting plants?" they demanded to know, then answered their own question. Men like Richard Ely were "visionary, theoretical dreamers," academic upstarts "posing as economists," meddlers who had "not a single day's practical experience providing electricity." And yet from their ivory tower they dared to suggest that private lighting companies were not public benefactors, and that government amateurs might wade into the business themselves and provide a better and cheaper light. Their economic treatises ignored "the actual experience of practical men and the best scientists of this generation, who have not only made the subject a study, but demonstrated the same by testing it every night for years."[38]

The private electric companies even denounced as un-American all those "hare-brained reformers" who criticized the private enterprise that was transforming the nation into a beacon of individual liberty and industrial creativity. Municipal ownership, the electric men charged, was a form of government "paternalism" unworthy of a free people and a hazardous step toward socialism. Richard Ely replied that the charge of paternalism was "a prize bogey with which to frighten the unthinking." As he put it, "the state is not something that is above us, doing something for us. It is one kind of co-operation. It shows greater self-reliance to provide a telegraph service for ourselves than to say, 'we are so dishonest and inefficient in government methods that we dare not trust ourselves. Will not some rich men kindly provide us with a good telegraph system, and please give us cheap rates?'" Declaring himself the true friend of democracy, Ely argued that those capitalists who put so little trust in democratic government were the ones who had "never become real Americans." In this way, the struggle over who should provide electric light became one skirmish in a wider Gilded Age war over the influence of large corporations on the economy, and a fight for the identity of America itself.[39]

Even at the height of the Progressive movement's drive for municipal ownership, local governments ran only a small fraction of the nation's electric utilities, but the threat did make private companies more amenable to a less drastic solution—what they began to call "reasonable" government regulation of their industry. For those towns and cities that continued to rely on private contracts to provide their lights, utility reformers such as Richard Ely proposed a range of ideas for improving the quality of service while holding the line on costs. In cities that adopted these reforms, companies sought contracts through an open bidding process. The winning firm enjoyed a temporary monopoly, shielded from the corrosive effects of cutthroat competition, and in turn agreed to open its books for review by a public utilities commission and its wiring to inspection by technical experts employed by the city. As one progressive economist put it, "publicity and public audit of accounts and inspection of the quality and safety of the service rendered by a monopoly are essential to intelligent action by the people."[40] If a company failed to deliver good light at a reasonable rate, reformers wanted citizens to have the power to rescind a contract by referendum.

Defending these new regulations, the progressive reformers pointed to Europe's example. Governments there had played a more active role in controlling the development of the electric industry. National laws encouraged municipal ownership of utilities and set standards for all electrical work, drawn up by leading scientists and engineers. And Europeans had done a better job of preventing the new technology from scarring their fine buildings and urban boulevards, placing most wires in the urban core underground early on. American progressives, many of them trained in German universities and familiar with London's example, could point to Europe as a model for the rational, public-spirited approach to providing electric light.[41]

But American electric companies drew a very different lesson from Europe's example. As they claimed time and again, the United States enjoyed far more electric light than any other place in the world, a bounty that electricians traced not only to their own inventiveness and entrepreneurial spirit but also to the lack of government interference. For American entrepreneurs, the slow pace of electrical progress in Britain seemed particularly instructive. Some blamed the aristocracy's "aesthetic conservatism" and the gas industry's inordinate wealth and political influence. Others pointed to the government's reaction to the reckless speculation in the first years of electrification. Eager to stake an early claim in what looked like a gold rush, many hapless British investors had ignored warnings from more sober "men of science" that electric lighting would take years and significant capital to perfect. Instead, as one Englishman put it, "stock-jobbers" took the reins away from engineers, and "the mad gallop that followed has ended in ruin and collapse." This tech bubble produced little electric light and much suspicion that the new industry had proven to be a failure. The innocent felt burned, as one put it, and "electric light suffered all the blame."[42]

Parliament responded by imposing strict regulations on electric lighting—price controls, inspection requirements, and safety standards that included the order to bury electric wires in cities. American electricians, and many of their British counterparts, blamed this government meddling for slowing the pace of electrification in England, imposing needless rules in an area best left to the free market. Even worse, British law discouraged private investment by giving local governments the right to buy out any lighting company's franchise after twenty-one years. As a result, British experts led the world in the science of electricity but the country lagged far behind the United States in actually installing lamps. True, America's lead in the field had been bought at a high price, as the unregulated market produced a number of accidents and fires that British engineers found horrifying. But big cities and small towns across America enjoyed the pleasures of an electrified present, while across Europe the electric light remained a thing of the future.[43]

Urban reformers were not the only ones pressuring America's electric industry to develop stricter safety standards. While utility companies might dismiss progressive economists as naïve and meddling communists who had no grasp of market realities, they could not so easily ignore the concerns of the fire insurance companies. When property owners began to install electric light in factories, mills, and stores, the insurance underwriters had to grapple with the fire risks posed by the new technology, and they used their influence to impose the electric industry's first safety standards.

The fire insurance industry had emerged as an important player in America's economic development at the close of the Civil War, when some of the largest firms agreed to form the National Board of Underwriters. This early "trust" tried without success to fix rates, but also worked to improve fire safety on a number of fronts. Serving the public interest and its own profits, this group lobbied cities to better train and equip their fire departments and improve their water systems. It drafted a model building code, pressured the police to take a harder line against arsonists, and encouraged the adoption of inventions such as sprinkler systems and electric fire alarms. Getting legislatures to pass fire safety laws proved slow and difficult but the insurers found they could "make virtue profitable" through market forces, offering lower rates to property owners who heeded these new rules and denying coverage to those who ignored them. In this way the National Board of Underwriters came to think of itself not only as a trade group protecting the private interests of insurance companies but also as a public service organization that championed any reform that might address the terrible urban problem of "fire waste."[44]

When factories and stores first installed lighting systems, insurance inspectors struggled to assess the risk this posed, handicapped by their total ignorance about electricity. Yet clearer safety guidelines became even more urgent as fires continued to destroy buildings. A third of the mills that installed electricity in those early years suffered fires, most minor but some catastrophic. Still, insurers found that developing an electrical safety code to guide their inspectors was no easy task. Rival electric companies guarded information about their products, and experienced electricians resented any meddling from the insurance companies' technical novices. In New England, some farsighted electric companies did recognize that cooperating with the insurance companies would be mutually beneficial. They formed an "Electrical Exchange," a forum that encouraged electricians to compare notes on the best practices in their trade, information that insurers used to draft some of the earliest safety guidelines for manufacturing and installations.[45]

But by the late 1880s, the need for stronger safety rules reached a crisis point. A number of insurance companies faced bankruptcy thanks to a "startling" increase in cataclysmic fires, including the one that destroyed so much of Boston. While this plague of urban fires had many causes, insurers believed that the chief culprit was electricity. The president of one insurance company declared, "We are standing in the presence of a mysterious element which no one is at present able to fathom," and insurers had powerful motivation to change that. The National Board of Underwriters gathered statistics on electrical fires and worked with leading electricians to draft national safety standards. For too long manufacturers of electrical equipment had each created their own standards, electricians followed no common practice in their installations, customers had no way of knowing if their products were safe and properly installed, and insurance inspectors trying to assess the risk of fire remained mostly in the dark, unable to distinguish safe electric systems from those on the verge of burning down a city block.[46]

The National Board issued illustrated manuals on the causes and prevention of electrical accidents, while insurance journals reported on the latest disasters. Many systems used wooden insulators that turned to kindling when wires overheated; textile factories hung lamps on pendant wires that accumulated grime and dust, then burst into flame; uninsulated wires exploded when the water from small plumbing leaks dripped through walls; electricians either failed to install safety fuses or used inadequate ones; and electric sparks continued to produce spectacular explosions when they ignited drifting clouds of leaking gas.

The electric companies never failed to point out that the actual percentage of fires caused by their product remained small; any fair tally showed that candles, oil, and gas posed much greater risks. But the burden on this new industry was higher since it sought to assure customers that it was safe to let this powerful and potentially deadly force into their factories, offices, streets, and homes. Whether or not the public's growing electrophobia was rational, the industry realized it would pose a significant barrier to expansion. To calm these anxieties, more electricians agreed to work with the "risk men" at the insurance companies. Through the 1890s, they developed safety rules and standards and upgraded their systems to reflect rapid improvements in insulation, fuses, and other safety devices.[47]

By the turn of the century, the electric industry's attempt to rationalize itself produced the National Conference on Standard Electrical Rules—the joint effort of insurers, electrical experts, architects, engineers, and manufacturers. Their "National Code" guided hundreds of towns and cities as they wrote the first electrical building and safety standards into law. Insurers also funded research at the Underwriters Laboratory (UL), a testing company that pioneered the field of consumer-product safety. UL was founded in 1893 by William Merrill, a young electrician trained at MIT who was hired by the insurance companies to assure the safety of lighting systems at the Chicago World's Fair. When the fair ended its yearlong run without electrical disaster, insurers continued to support Merrill's work testing the safety of various wires, switches, lamps, and appliances and tracking the causes of electrical fires. For the first time, the fire risks posed by electrical devices received careful scientific scrutiny. In the early twentieth century, UL evolved into a nationwide organization, with a fireproof laboratory in Chicago and branch offices around the country. The first of its kind in the world, the lab offered manufacturers a safety certification that assured their customers that an electrical product had passed a rigorous independent test and was unlikely to be an agent of destruction when brought into the home.[48]

Urban reformers, insurance salesmen, and utility companies agreed on the need not only for clear guidelines for electrical safety but also for better electricians. Soon after the 1889 Boston fire, MIT president Francis Walker told Americans they had good reason to be "terrified" about the dangers of electricity since so few people working in the field had the proper mix of scientific training and practical experience that it took to run an electrical system safely. Walker divided most of the first generation of electrical workers into two types, "cranks full of enthusiasm and theory" and blundering workmen with hands-on experience but no grasp of the first principles of electricity. In the wake of the public backlash against electricity in the late 1880s, many industry leaders agreed that the bright future of their business could only be realized with the help of a new breed of professional electrical workers, properly trained in both the latest scientific theory and the practical challenges of building and operating power stations. MIT, Cornell, Columbia, and Harvard responded by establishing some of the first graduate programs in electrical engineering. When Edison had launched his first power station in 1882, no school in the country taught electrical engineering; he had been forced to create his own company schools to train his employees. Twenty years later, forty-nine colleges around the country offered a four-year degree in the field, serving over two thousand students.[49]

This first generation of university-trained engineers joined an industry that was being transformed by corporate mergers, one part of a broader movement toward economic concentration and rationalization in the 1890s. In 1892, J. P. Morgan orchestrated the creation of General Electric, a heavily capitalized corporation that combined many of the Edison Company's assets and patent holdings with the more savvy management approach of the Thomson-Houston Company, one of the inventor's major rivals and an early leader in AC power. Managed by Thomson-Houston's president, Charles Coffin, General Electric soon controlled well over half of the market in electrical manufacturing. Most smaller companies not swallowed up by GE were absorbed instead by Westinghouse, and in 1895 the two corporations agreed to avoid further costly litigation by pooling many of their overlapping patents. The various manufacturing enterprises produced by the first generation of electrical pioneers evolved by 1900 into a duopoly of "electric trusts" that controlled the development of electrical technology through the twentieth century.[50]

The General Electric merger marginalized Edison in the very industry he had done so much to create. But GE provided him with a payoff that afforded some relief from the years of financial stress he had experienced in trying to bring his lighting system to market. As usual, Edison invested that money in further invention. By then he had replaced Menlo Park with a new "invention factory," an extensive laboratory in Orange, New Jersey, designed to be "incomparably superior to any other for rapid & cheap development of an invention & working it up into commercial shape." In addition to conducting experiments in everything from ore separation to the bleaching of cigar wrappers, Edison's research team experimented with X-rays, improved his phonograph and moving picture machine, and continued to refine his direct-current electric light system.[51]

Over time, the corporate mergers of the 1890s transformed the invention process in the field of electricity. Building on the team approach Edison had pioneered at Menlo Park, General Electric and Westinghouse went into the research and development business for themselves, founding their own laboratories staffed by talented graduates from the new university programs in science and engineering. By the early twentieth century, these teams of university-trained specialists guided progress in the field of electric light and power, working in tandem with their company's manufacturers and sales teams. The ingenious mechanic and the entrepreneurial adventurer yielded to credentialed professionals, their work directed not by their own inventive muse but by the market strategies of their corporate managers.[52]

Even before this happened, the electrical revolution that Edison had done so much to create surged past him. By the early 1890s, scientists and engineers with a stronger grasp of science and mathematics took the industry in directions that the great inventor found uncongenial, perhaps even incomprehensible. Convinced that the high-voltage alternating-current generators produced by his rivals would prove both inefficient and too dangerous, Edison had squandered his lead in the lighting field by sticking with his direct-current system. As the scientific consensus moved in favor of AC power, taking many customers with it, Edison tried to win this "battle of the currents" by appealing instead to the court of public opinion, playing on public fears about the dangers of high-voltage wires. To make the point most vividly, one of Edison's allies invited reporters to watch the electrocution of a series of hapless stray dogs and farm animals, using one of George Westinghouse's AC dynamos. In his eagerness to demonize alternating current, Edison wasted some of his great scientific credibility by encouraging New York's decision to use the electric chair to execute a convicted murderer.[53]

Now middle-aged, Edison had come to play the unusual role of conservative skeptic who mostly looked on as his rival inventors solved the many technical challenges posed by AC power. Others created the dynamos, transmission lines, transformers, meters, and safety devices for the high-voltage AC system, taking the next crucial step toward the creation of the modern electrical grid. Most notably, the brilliant Nikola Tesla, a former Edison employee, pioneered the development of the AC motor, and amazed audiences by demonstrating an early version of fluorescent bulbs that dispensed with a filament altogether. A decade after launching the Pearl Street station in Manhattan and winning the race to create a viable incandescent light, Edison felt increasingly irrelevant in the industry he had done so much to found, confiding to his secretary that "I've come to the conclusion that I never did know anything about" electricity. In 1897, he sold his remaining shares of the Edison Electric Illuminating Company of New York and declared his intention to retire from the electric lighting field.[54]

As the electrical industry matured by the end of the nineteenth century, its managers, manufacturers, and engineers established themselves as professionals, one part of a wider process of economic rationalization that transformed corporate capitalism in these years. They credentialed themselves at universities and joined professional organizations that lobbied to protect their interests. But those doing the hard and dangerous work of stringing and maintaining electric wires enjoyed no such status. Rather, they came from what one called "the floating population—men having no trade or steady occupation, and ready to take a 'hack' at anything that came along." Some admired the linemen, who worked like sailors aloft in the electrical rigging, high above the surging sea of pedestrians. But others, including some employers, expressed contempt for these men, many of them transients who worked for about two dollars a day in an occupation that suffered an accident and mortality rate twice the national average.[55]

Experienced linemen felt threatened not only by dangerous live wires but by the relentless deskilling of their occupation. "In the blind, senseless competition for work," as these veterans saw it, "cheapness has almost become the prevalent rule, to the detriment alike of employers and journeymen, to the injury and danger of the public, and to the ruin and degradation of our trade." In the race to keep up with demand, the companies had hired many raw recruits, men brave or desperate enough to climb poles and string wire but ignorant of basic safety practices. The old hands called these newcomers "butchers in the business" because their shoddy work accounted for many of the electrocutions and fires that plagued American cities.[56]

Like other workers in these years of labor strife, linemen turned to union organizing as a way to protect themselves and their livelihood. "A growing number of self-educated wiremen," as one put it, "are beginning to feel the dignity of their profession." Some joined the Knights of Labor, others formed an uneasy alliance with organized telegraph workers, and in 1891 a small group laid the groundwork for what became the International Brotherhood of Electrical Workers. Like any other union, the group organized to support workers injured on the job and paid a death benefit to the widows of fallen comrades. Struggling through the depression years of the 1890s, the union won some strikes for better wages but lost many more, including a bitter conflict with the Edison Company.

Union workers agreed with their employers on a single point: something had to be done to improve the standard of workmanship in their field. At a time when the technology was rapidly evolving, union locals invested in blackboards and reading rooms offering the latest scientific journals and organized classes and lectures to help workers keep up with the latest developments. Over time the union would push for licensing requirements for their trade and stricter safety regulations, doing all in their power to ensure that no other electrical worker suffered the terrible fate of lineman John Feeks.[57]

By the turn of the twentieth century, electricity was coming of age, no longer a curiosity but a mass commodity, delivered by a sophisticated and heavily capitalized industry that saw nothing but exponential growth ahead. Some old-timers who gathered at the National Electric Light Association conventions looked back nostalgically on the early years when any man with a bit of enterprise, practical skill, and a fascination with electricity could go into the light business for himself. But most acknowledged that the changes had been necessary, the only way for industry to grow, providing more, safer, and less expensive light while paying healthy dividends for its investors.

Though the rash of fires and electrocutions had threatened the industry's growth in the late 1880s, it emerged from these challenges stronger than ever. The owners and operators of the utility companies deserved a share of the credit for turning electric light from an invention into a paying business, but the expanding electrical grid was also created by other institutions and conflicting forces, an attempt by many Americans to mediate the interests of rival inventors and manufacturers, the salesmen of light and their customers, the competing values of free enterprise and progressives' call to use government to protect the public from the twin dangers of market chaos and corporate monopoly.

The more mature and stable electrical system that emerged by the early twentieth century had been produced by other forces as well—the scientists and journalists who helped to develop and spread a common language of electricity; research institutions and technical schools that turned a growing number of enthusiastic students into licensed electricians and engineers; legislatures that

Over time, cities built what reformers called the "scientific street." While traffic passed above on paved roads, conduits below ground housed not only the new subway train but also separate pipes for sewage, water, gas, steam, and various electrical wires.

. . . .

created regulatory bodies and electrical codes; the experts who protected the public interest through their work as inspectors, utilities commissioners, and civic-minded economists; unions that tried to protect their members from unsafe work practices; and the insurance companies that guarded the public's interest and their own bottom line by imposing safety standards on electrical products and work. All played their part in turning Edison's famed invention into the far more complex and powerful creation, the modern electrical grid.

As electricity came increasingly under the control of corporate managers, university-trained specialists, and accredited technicians, it grew ever more opaque to the average American. Like any other group of professionals, electricians came to speak an inscrutable jargon, "a mystery to all but the initiated." Even the basic units of their trade—"ohms" and "volts" and "watts"—were the alphabet of an alien language to most Americans. Electricians often joked about the public's "wild views" about electricity, but some in the industry considered this ignorance no laughing matter. Journalists and politicians with half-baked ideas about electricity were bound to demand the impossible from electric companies and stir up public fear and resentment against their business. Better education seemed the answer, and many industry leaders called not only for technical training for electrical workers and engineers but the addition of electrical science in the high school curriculum. The public's irrational fear of electric power was bound to fade, these men hoped, when "wider education" gave more Americans a chance to explore and understand electricity for themselves.[58]

The industry had much to gain, of course, from what it called this "high mission," since it would cultivate young people eager and able to contribute to the growth of the business. By the turn of the century, various electrical trades employed nearly a million peo-

"THE GENIUS OF LIGHT" STATUE.

When Edison saw The Genius of Light *in Paris, he bought the life-size marble statue for his library in Orange, New Jersey. The figure perches on a broken gas lamp and holds aloft a working Edison bulb.*

....

ple, making electric light and power one of the nation's largest and fastest-growing industries. Progressive educators did not care so much about producing good employees for the electric companies, but they agreed that every American student should receive at least the rudiments of a technical education. Consumers had to know enough to make informed choices, they reasoned, and voters needed to understand technical issues in a world where power was both provided and wielded by utilities companies.[59]

But the democratization of electrical knowledge was an uphill battle. As the industry grew larger and more specialized, and as the science became more abstract and demanding, the gulf between the technical experts and the general public widened. Most citizens only knew electricity to be a "mysterious force," one electrician complained, "coming from nowhere in particular, doing very nearly what it pleases, and equally able to produce a spark and a general disruption of the universe."[60]

In spite of its "very dense ignorance" about electricity, the public remained fascinated by the subject. Long after the first flush of excitement about electric light had worn off, popular journals still slaked a thirst for news about the latest electrical conveniences and predictions about the ease and glamour of an electrified future. Lectures on advances in electricity continued to draw large crowds, though often the most successful ones pandered to the public's interest in novelty rather than insight. For example, a lecture in Boston left the audience howling with delight over tunes played on an electric piano and a séance featuring glowing skulls dancing the cancan. In short, few could fathom what electricity *was*—but this never stopped them from enjoying its mysterious pleasures and indulging in wild speculation about what it *meant*. By the close of the nineteenth century, most Americans decided that it meant they were living in the most wonderful civilization that the world had ever known.

Nine

The Light of Civilization

On a steamboat tour down the Mississippi in 1882, Mark Twain marveled at the way birds reacted each night when the ship's spotlight lingered on the riverbank. Some of them "tuned up to singing," while hundreds more flocked out of the trees, "careening hither and thither through the white rays." For many creatures, artificial light's power to upend the timeless rhythm of day and night proved even more disorienting than it did for humans.[1]

The new light's "curious effect" on animals had a tragic side, as reports soon circulated about the mass destruction of birds, lured to their death by the new technology. In Des Moines, "urchins" gathered under the wires every morning to scoop up "hatfuls" of sparrows electrocuted when the light company turned on the power each night. Migrating ducks and geese often fell prey, disoriented by arc lights mounted on towers and tall buildings. In Cleveland, a passing flock collided with a series of tall arcs, smashing the globes and extinguishing half of them. Below, amid the shards of broken globes, people gathered the still warm bodies and "congratulated themselves upon the novel and inexpensive method by which they had secured their game." Heaps of bird and bat carcasses piled up at the base of the new Statue of Liberty, leading some to call for its torch to be extinguished. Instead, attendants put the most interesting specimens on display and tossed the rest into the harbor. A steamship on the James River, "ablaze" with electric light, lured more than a hundred canvasback ducks into a fatal collision. "Country ducks," the reporter joked, "are not familiar with Edison's invention." One of the worst incidents occurred when migrating flocks were drawn to their death by the powerful arc lights atop Chicago's Board of Trade. A watchman the next morning found the building's roof covered with the colorful carcasses of dead or dying birds, many of unknown species, and the sidewalks below were thick with enough "to trim all the ladies' hats in Illinois."[2]

One bird electrocution proved so sensational that newspapers around the country picked up the story. Near Fresno, California, two eagles touched while both sat on a ten-thousand-volt transmission wire, making a connection that incinerated them in an instant. All that remained for the linesmen to discover were two talons "burned to a crisp," still clinging to the wire above, and a skull and scattered parts below. Some papers treated their readers to an artist's rendition of the eagles' electrocution and photographic reproductions of their charred remains.[3]

Many species of insects also proved vulnerable, unable to resist flying to a fiery death or banging themselves to exhaustion against the globe of a streetlamp. When the federal government experimented with arcs on the Capitol and other public buildings on the National Mall, "billions" of dead insects quickly piled up in nooks and crannies of each building, while thousands more clung in a "death grip" to the walls. One scientist estimated that a single arc light killed one hundred thousand bugs each night. Since some species respond particularly to blue light, the arc's glare proved especially tempting. H. L. Mencken reminisced that when these lights first came to Baltimore in the mid-1880s, the lamps drew a species of large water beetle in swarms so thick that the light itself was obscured. Each night thousands crashed to the sidewalks below and were crunched underfoot by pedestrians. Mencken's pals were not the only ones who came to the medieval conclusion that these "electric light beetles" were spawned by the arc lamps themselves. The manager of a St. Louis electric plant had to assure his anxious customers that the creatures had "always been here, only they were never visible until the electric light was introduced." The rotting carcasses stank, and, making a bad situation worse, rumors spread that the bite of an electric light bug was "dangerous as a tarantula's." One professor conceded that walking past an electric light had become a "great nuisance," but concluded that this was "one of the evils which came with the electric light and we have to put up with it."[4]

Wherever lights went on, they served a second function as an exterminator of what a journalist called all those "ugly little bodies." Farmers erected arc lights on the outskirts of their fields, hoping to lure insects away from their crops. In Louisiana, some dreamed of controlling mosquitoes by illuminating the swamps, though they soon learned that while the harmless males found the dynamo's drone to be irresistible, the females that cause so much human misery were unmoved by the light. In the end, most abandoned the use of light to control bugs, since it proved better at attracting than eliminating them.

Some saw in these curious incidents of mass destruction a great opportunity for scientific study. No need to venture into the wild when one could find hundreds of rare species just by sifting through the mounds beneath an arc light each morning. Smithsonian scientists collected bird specimens at the foot of the Washington Monument, hoping to learn what they could about migratory patterns of night-flying birds. These scientists regretted the "slaughter" caused by this collision between animal instinct and human improvements, but most observers expressed more curiosity than concern over the massive destruction of birds, bats, and bugs. Many complained about the sight and smell of heaping piles of insect carcasses but were otherwise enthusiastic about the light's unexpected power over the animal world. Here was a fine new weapon, they thought, in humanity's ongoing attempt to subdue and shape nature.

At a time when American market hunters were driving the passenger pigeon to extinction and the bison and wolf nearly to the same fate, they embraced the advantage that strong light provided them over game animals that were drawn and dazzled by its beam. Among the first to report success was a pair of hunters from Massachusetts who traveled to Louisiana, rented a portable arc light, and turned it on a canebrake. Over the course of two months of night hunting they flushed out thousands of game birds, mostly woodcocks—a valuable delicacy in those days. "By tens and twenties," one of these "sportsmen" reported, the woodcocks hurled themselves against the light. Only fifteen yards away, fluttering in the "superb artificial day," the small birds looked "big as barns. . . . Whang, whang went the guns, and down they would come, fit morsels for a king." The pair killed more than three thousand birds, shipping them to markets from New Orleans to Chicago. A single glitch came when resentful locals fired a shot through their light, causing damage that could be repaired only by offering the electrician a large tip and a stiff drink to steady the poor man's nerves. "Why people should be so peculiar over a few birds I cannot understand," the sportsman wondered. "They ought to have admired our Yankee ingenuity and enterprise."[5]

Likewise, fishermen recognized light's power as a lure. Sportsmen rigged battery-powered fishing lures that illuminated tiny "pea" bulbs, while commercial fishermen experimented with ways to rig incandescent bulbs in their nets. "The hauls are immense," one paper reported, "so much so that in the case of the salmon fishery there is a danger of overdoing it, and of rendering the species extinct. This would be as bad as killing the hen that laid the golden egg."[6]

Larger game, deer and elk, also proved vulnerable to electro-hunting. Hunters rigged "jacklights"—small but powerful head-lamps or even larger spotlights mounted on rowing skiffs. As one hunter explained, "When the dazzling rays of a jacklight strike a deer, it turns and faces it, startled for the moment into stupid curiosity." Many denounced the practice as "utterly unsportsmanlike" and pushed through state laws banning jacklighting, though poachers continued the practice. Completing the circle, electricians rigged small electric bulbs on the antlers of stuffed deer and elk, creating "novel electroliers" that hung in saloons and were much admired. Victorians on both sides of the Atlantic enjoyed turning the heads, feet, and stuffed carcasses of exotic animals into lamps and other furniture, and lost no time in rigging these prizes with incandescent bulbs. Few seemed troubled by this juxtaposition of human comfort and animal degradation—both only confirmed that humanity not only held the highest point on earth's chain of being but, through its mastery of technology, had climbed a few steps closer to the angels.[7]

In this age when so many Americans felt a palpable pride in their society's great technological inventiveness, the first flip of an electric switch was often observed with ceremonial pomp. None enjoyed the theatrical possibilities more than the Businessmen's Club of Evanston, Illinois. In 1890, it inaugurated the use of electric lights in its hall by staging a pageant of progress. As the band played a dirge, the old gaslights gradually dimmed and went out; once the audience had been submerged in darkness, the music switched to a jubilant march that offered a rousing welcome to a brighter future, in the form of the "soft radiating light" from three hundred incandescent bulbs. In similar ways, people across the country honored their first moment of electrical illumination as a historic event, a tangible moment when they left the dim past behind and stepped into an electrified future.

Electric light formed the centerpiece of a grander pageant of progress, the epic Broadway show *Excelsior,* which played to packed houses at Niblo's Garden and later toured the country. Directed by the Hungarian immigrant Imre Kiralfy, the same man who orchestrated *The Fall of Babylon* on Staten Island, *Excelsior* not only pioneered the use of electric light on Broadway but also cast the new technology as a central character. Combining pantomime, ballet, and "magnificent tableaux," the show portrayed a series of struggles between Darkness and Light, symbolic of "the strife betwixt knowledge and ignorance which has constantly attended the advance of civilization." Hundreds of spangled dancers in sea green and salmon-colored togas cavorted among the spotlights, leading to a climactic struggle at the "Temple of Light." There, Light banished the dark shadows of superstition and slavery once and for all, with an effulgence of colored beams that one reviewer hailed as "indescribably gorgeous."[8]

As Kiralfy recognized, the light's symbolic power was best appreciated when contrasted with the surrounding darkness, a principle that seemed as relevant to the Gilded Age student of culture as it did to the stage producer and the lighting engineer. Most social critics, civic-minded preachers, and prognosticators of the future offered their audiences hearty congratulations for their good fortune to be born into an era when the forces of light were winning an unprecedented triumph over the dark. One only needed to conjure

Edison personally supervised the installation of the five hundred lights that helped to make Excelsior *a Broadway "spectacular," and took a keen interest in the play's novel lighting effects.*

. . . .

what life had been like for those elders who could still tell stories about their first encounter with the telegraph and the steam engine, once marvelous but now ubiquitous.

While many adults could remember this dimmer pre-electric past, late-nineteenth-century intellectuals often framed their era's technological revolution in a wider context, placing it in the long arc of human history. Advances in geology, astronomy, and the biological sciences inspired scholars on both sides of the Atlantic to adopt a more expansive view of time, reflected in a new interest in both historical method and evolutionary theory. Under this influence, they embraced the nineteenth century's revolutionary improvements in artificial lighting as an ideal metaphor for the wider progress of human civilization. Just as Kiralfy had done in his Broadway pageant, these writers understood human history as an epic journey out of darkness, as the shadows of want, ignorance, and injustice were slowly but surely dispelled by humanity's growing technological power and intellectual enlightenment.

And so, many late-nineteenth-century academics produced popular articles and quasi-scientific treatises on the history and evolution of artificial lighting, each depicting the light bulb as the fulfillment of humanity's quest to conquer the dark. Ancient humans, these works explained, knew only the smoky smudge of a light—a shell or animal skull with a simple wick that flickered over a fetid pool of fish oil or animal fat. For most of human history, the simple tallow lamp had been all the light most had ever known, but the scientific and technical breakthroughs of the nineteenth century had produced a rapid succession of ever more sophisticated techniques for satisfying mankind's craving for more light.[9]

As these historians often explained, the evolution of lighting systems, from candle through whale oil, kerosene, coal gas, and finally to arcs and incandescence, mirrored the wider growth of civilization and its mastery of nature. Invariably, they agreed that stronger lighting pushed back not only the physical darkness but also a spiritual one. Illuminating the higher reaches of the soul, each new form of artificial light produced corresponding improvements in civilization's morality and its appreciation of beauty. In this way, the quality of a culture's artificial lighting revealed its level of civilization, and one need not look to the distant past to see that primitive cultures used primitive lamps. Like living fossils of mankind's ancient history, late-nineteenth-century Greenlanders still relied on whale blubber, the "southern Negro" his pine torch, and the "Red Indian" his "rude camp fire." Some Polynesian tribes were said to have "no lamps whatsoever," a developmental stage of "simple savagery" that most races had left behind "ten thousand years ago."[10]

At a time when white Americans were constructing the apartheid walls of Jim Crow, newspapers relished stories that denied African Americans any part in their country's technological accomplishments. In a number of fields, including electric lighting, black inventors made significant contributions despite their obvious disadvantages in education and opportunity. The mainstream press ignored these, preferring instead to publish tales that made African Americans seem as bewildered as moths when they first encountered the new light. Thus an Albuquerque paper shared the story of a "colored gentleman" who was the last one in his town to understand that he could not blow out a light bulb, a feat he tried "until his eyes bulged and the sweat trickled in rivulets from his features." In another popular variation, a "negro servant" assured his white boss that he understood how the master's new lighting system worked. "I un'stands all 'bout dem dynamos and pow'-houses an' sich," he explained, "but how do the kerosene squirt thoo 'dem wiehs?" As if to make the same point in the most gruesome way possible, some lynch mobs used the masts of the new electric light poles to hang their victims, proof of their shamelessness and an apt symbol of their assumption that electricity was the white man's badge of cultural superiority.[11]

Across the globe, nonwhite people's first reaction on seeing electric light provided modern readers with condescending insights into "the native mind." Indians in Bolivia, for example, reportedly tore down poles and wires, convinced that the electric light was trying to "swallow the moon." A Western visitor to Tehran reported that superstitious Muslims denounced that city's first arc light as the work of satanic forces. Perhaps the most popular and often reprinted tale involved the British army's use of spotlights as a weapon against its "barbarian" enemies in the Sudan, repulsing a nighttime attack with a powerful beam that confused and scattered a "howling, rushing mass of Arabs."[12]

In the late nineteenth century, Christian missionaries brought electric light to their remote posts around the globe—a powerful tool that seemed to vindicate both their benign intentions and Western Christendom's cultural superiority. A dynamo's mysterious power was sure to "befuddle the native medicine men" and undermine their authority, while the useful beauty of its light would confirm, by association, the Christians' spiritual "lamp of truth." When the first missionaries announced plans to carry a portable lighting system into the "the remotest recesses of the Dark Continent," their supporters applauded this attempt to throw electricity's "genial beams upon the dusky natives," one part of a wider mission to promote morality and civilization. Ever on the lookout for new markets for electricity, the *Electrical World* was quick to note, "No missionary steamer is fully equipped for its work without an electric light plant."[13]

When westerners could not bring electric light to the "savages," they eagerly brought savages to the electric light. Pioneers in the new science of anthropology "discovered" many isolated tribes in these years, such as the cliff-dwelling Tarahumara people of Mexico. When an American explorer convinced seven of them to leave their remote mountain homes for a train trip to Chicago, reporters eagerly followed the story. These "genuine savages" gave American readers full satisfaction by reacting to modern technology "with emotion akin to abject terror." First approaching a city ablaze with light, they "trembled mightily" at the sight of this "mysterious illumination." Repeated many times across the globe, such encounters between traditional cultures and electric light offered readers a racial framework to understand the unsettling changes wrought by their technological revolution. As scientists assured the American public, these primitive people were living fossils, mental children locked in "the first stage of human evolution." Vicariously looking at the latest inventions through the eyes of these "savage races," modern people could better understand and appreciate how far their own society had advanced, and take pride in their privileged place at the forward edge of human development.

Such parables satisfied much more than idle curiosity and "scientific interest." At a time when Europeans and Americans spanned the globe to win new markets and impose imperial power, and worried about the impact of immigration on their own hereditary stock, these stories affirmed the white man's right to rule at home and abroad—not as conquerors but as benefactors and vanguards of the human race. When Europeans first began exploring the wider world in the fifteenth century, they considered Christianity to be the source of their cultural superiority. In the midst of the Industrial Revolution of the nineteenth century, they pointed as often to their advanced science and their more powerful machines. As the American imperialist Josiah Strong put the case, "Nothing so well illustrates Man's triumph over nature and his control over physical conditions of life as invention, and in this sphere the Anglo-Saxon has no rival."[14]

Those who considered electric lighting a distinctive mark of European or Anglo-Saxon cultural superiority puzzled over the anomaly of Japan. Western intellectuals often claimed that the Asian races "lacked progressiveness," but after the Japanese ended centuries of self-imposed isolation in the late 1860s, they embraced Western technology with an enthusiasm and skill that many found "astonishing." Paying them his highest compliment, one American visitor described the Japanese as "like the Americans in their ready adoption of new things. . . . They are quick-witted, and want to be up with the times." Their government installed a Brush arc system in the imperial palace in 1887 and sent envoys to the United States and Europe to study the latest technological advances. When word reached New York that eight hundred Japanese men had joined their country's first electrical club, the *Electrical World* confessed that it "hardly knows how to receive this piece of intelligence."[15]

American companies enjoyed a lead role in the electric light and power business across the globe, but some warned that the Japanese would soon turn from lucrative customers into dangerous competitors. Many Western observers still judged them to be an "imitative race," unlikely to generate new ideas on their own. But none doubted their skill as manufacturers. "The Japanese can make anything we can," as one editor warned a decade after electric light's arrival in Japan. "They can produce what we need at less cost than we can make it ourselves, and unless a high protective tariff is raised against Asia, that country will become the factory for America." Unsettling the Western assumption that modern technology was an expression of Anglo-Saxon racial superiority, the Japanese earned begrudging admiration as "the Yankees of Asia." Carrying this idea to its biological conclusion, a U.S. government report attributed Japan's growing industrial prowess to a trace of "Aryan" blood in the nation's hereditary stock, described in the same report as an "undoubted white strain."[16]

. . .

The writers of pulp-fiction novels in the late nineteenth century embraced this link between racial hierarchy and technological superiority, none more than Luis Senarens, who wrote a couple hundred stories about the intrepid boy inventor Frank Reade Jr. and his adventures among the exotic colored races of the world. Most of his tales opened with young Frank, "the greatest of living inventors," unveiling a new and ingenious machine, often to an astonished and fawning reporter. Steam power was the latest marvel when the series began in the 1860s, and so the early volumes offered readers the exploits of his "steam-powered horse" and "steam-powered man." By the 1880s, both Reade and his readers had moved from steam to electricity, and the inventor produced an electric horse, various electric airships and submarines, and a powerful but obedient servant, the eight-foot-tall Electric Man.[17]

The author of these tales wasted no time attempting to explore the science behind Reade's implausible inventions; the gleam of confident intelligence in the handsome young inventor's eye seemed sufficient explanation for the success of an "electric air canoe" or an "electric ice-boat." As dazzling as these contraptions might be, none could hold a reader's interest for long while idling in the inventor's workshop, and each soon played a lead role in another amazing adventure. Accompanied by his "faithful attendants" Pomp and Barney, an affable "darkey" and a pugnacious Irishman, young Reade took his inventions to exotic locales where he tested the machine's mettle against both the forces of Nature and the ignorant and often malicious hordes of less advanced races. In hundreds of installments, Reade's inventions baffled Apaches and Amazonians, cutthroat Bedouins and Chinese bandits—rivals no less menacing and exotic than the sea serpents and man-killing tigers that Reade also conquered with his superior technology.

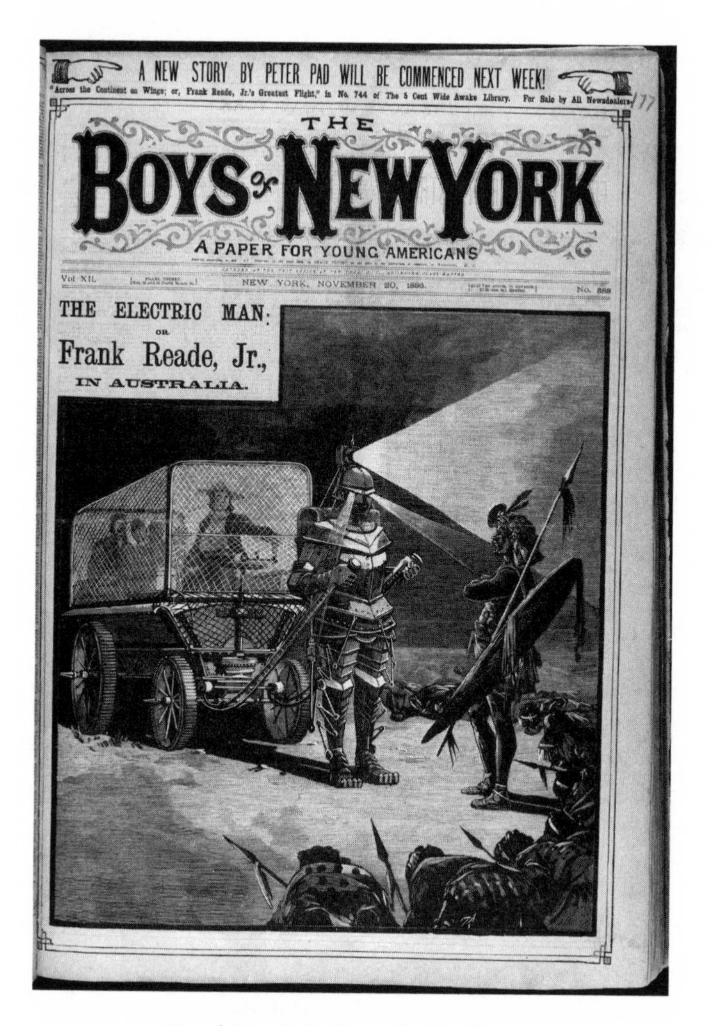

Frank Reade Jr. brought his Electric Man to the Australian outback.

....

Typical of this genre was the first encounter between Frank Reade's Electric Man and a ragged band of Australia's "native and wicked bushmen." The boy inventor and his ethnic companions brought the Electric Man to Australia for a scientific expedition to uncover the mysteries of that continent's vast interior. Reade's fame had preceded him to Sydney, where the residents greeted him with a lavish banquet. "Though an American he belongs to the world," the mayor declared. "He overcomes the elements—he penetrates space and rescues the unfortunate from the jaws of death!" The city's beautiful women could not have agreed more, and vied for a chance to meet the handsome but modest American genius.[18]

Soon the Electric Man towed a specially designed wagon that carried Reade and his assistants into the Australian bush, where they shot a kangaroo, survived an earthquake, and discovered gold, all the while facing the threat of attacks from "savages." On the first night a band of aborigines encountered the Electric Man, they quite understandably stared in wonder, and then trembled and howled once exposed by the powerful beam of the robot's spotlight. "The glare of the light was too much for them," it seemed, "and down they went on their knees." All except their "stalwart" chief, who stared defiantly into the light, wearing an "expression of mingled wonder, fierceness and surprise." His followers could only grunt, "Ugh! Ugh! Ugh!," but with more courage than sense the brave chief flung spears at this incarnation of American ingenuity, only to see them shatter to the ground. Reade dispatched the persistent primitive with the turn of a crank that caused Electric Man to deliver a mighty metallic kick to the chief's stomach, making him "the sickest man ever seen in Australia."[19]

In many of the Frank Reade stories, the author exploited the electric light for all of its narrative potential. The boy inventor and his sidekicks found themselves in one dark spot after another—stormy nights, moonless savannas and jungles, or the even blacker worlds of caverns and the ocean floor. Creatures that would be menacing under any circumstance seemed even more so when they burst out of the dark into the brilliant beam of Frank Reade's spotlights. Attached to diving suits and the bows of air canoes, these lights were often described as "the most powerful ever invented," and never failed to expose danger, reveal treasure, and rout superstitious natives. Light's power to startle as it plucked objects out of a dangerous gloom thrilled the readers of Gilded Age pulp fiction, a literary invention that has been put to work ever since in the genres of horror, mystery, and science fiction.

The link between technological progress and Western cultural superiority found its apotheosis in the 1901 Pan-American Exposition in Buffalo, New York. Since the Chicago World's Fair eight years earlier, every major exposition had included elaborate light shows, but Buffalo was the first to make electric light and power its central theme. Using electricity produced by the "enslavement" of the nearby Niagara River, the fair burned more bulbs than ever before, offering visitors "an Incomparable Vision of Illuminated Loveliness." While the Chicago fair had used nothing but pure white light, Buffalo constructed a "Rainbow City" of exuberant but carefully orchestrated colors, a bold experiment in the new language of light.[20]

Fair planners installed this incandescent fantasia not only to amuse but also to teach. The architecture, lighting, and color scheme of each building were designed to give visitors the experience of walking through time, tracing the heroic ascent of human evolution. The popular midway at the 1893 World's Fair had offered a glimpse of the exotic and often erotic world of "primitive" cultures, but organizers had intentionally kept these lowbrow attractions outside the fairgrounds. The planners in Buffalo chose instead to incorporate the midway's exhibits of primitive cultures into the exposition, assuring visitors that they would find them both amusing and "instructive," scientifically arranged to offer a vicarious encounter with every stage of human evolution.

Most visitors to the Pan-American Exposition entered through the midway, where they encountered wild animal shows—a diving elk, an "Educated Horse," and Esau the "Missing Link," a chimpanzee who could play the piano and bang on a typewriter. Close by these animal attractions, visitors found exhibits devoted to the exotic but "lower" branches of the human race, offering "curious and interesting evidences of civilization, so different from our own." Fairgoers could peer into "Darkest Africa" or enjoy the sight of 150 "Southern Darkies" doing "a grand cake walk" and other "plantation songs and dances." They could also observe an Indian Congress that gathered hundreds of "real natives" from dozens of tribes, watch Eskimos dancing in an ice grotto, see Mexican bullfights, and enjoy camel rides and suggestive oriental dancers in a "Moorish Palace." These exhibits of nonwhite cultures were painted in the "strongest, crudest colors," warm earthen tones meant to suggest the primal passions and childlike sensibilities of those races still dawdling at the starting gate of human progress.[21]

As visitors left the midway and moved into the fair, the color scheme of the buildings grew subtler and ever lighter, symbolic of Western man's ascent to greater enlightenment. Over the "Bridge of Triumph" and past the "Fountain of Abundance," fairgoers moved toward the literal and symbolic center of the grounds, a grand Electric Tower—"the altar at the head of the aisle of increasing brilliance." Lit by forty thousand bulbs, washed with sprays of colored spotlights, and crowned by a gilded statue of the Goddess of Light, the tower was billed as "the most marvelous achievement of the age of its kind." "This is a City of Living Light," a reporter declared, "and those who look upon it know that in their wildest flights of fancy they never conceived anything to which it can be compared." Even Thomas Edison was impressed, declaring it "the apotheosis of incandescent light."[22]

Historian Robert Rydell describes the Pan-American Exposition as "a carefully crafted allegory of America's rise to the apex of

The Pan-American Exposition's Electric Light Tower epitomized what some called "nocturnal architecture," a new kind of building designed to look best after dark. "It shines like diamonds," one visitor exclaimed. "It is like a transparent soft structure of sunlight held stationary against the background of darkness."

civilization." Close by the Electric Tower, brilliantly illuminated exhibit halls offered visitors a chance to review the very latest achievements in the mechanical and electrical arts, expressions of "the genius of the American people." Many demonstrated the steady improvements being made in fields already known—more efficient lamps and motors, and Edison's latest invention, an improved storage battery for use in electric automobiles. But visitors also crowded around demonstrations of the newly discovered and mysterious X-rays, the light-shedding element radium, and wireless telegraphy. Unsettling what scientists thought they knew about the nature of matter, these marvelous discoveries suggested that the quest to know and master the force of electricity was yet in its infancy.[23]

Hours before being felled by an assassin's bullet, President William McKinley spoke at the Buffalo fair, describing such expositions as "the time-keepers of progress. . . . They stimulate the energy, enterprise and intellect of the people and quicken human genius." But another reviewer spoke for most fair visitors when he wrote that the display of new inventions and strange electrical forces "stirs our pulses with pride, even if we gape at it with unenlightened eyes." New technologies and the science underpinning them grew ever more central to the lives of early-twentieth-century Americans, and ever more incomprehensible. However, the racial hierarchy built into the exposition's very design and color scheme offered white visitors the comforting suggestion that they were apparently *less* confused by technology than those "less civilized" races left behind on the fair's midway. As whites of European stock, even the technologically naïve and bewildered could take pride in belonging to the inventive race that served as the vanguard of progress.[24]

Not all accepted this technological justification for racial hierarchy and Western imperialism. Among the sharpest critics was Leo Tolstoy, the Russian novelist and Christian ascetic who inspired an entire generation of religious radicals and pacifists. "Let us not deceive ourselves," Tolstoy wrote, as he tried to smash what he considered to be one of the great false idols of his age. Inventors care nothing about the welfare of humanity, he argued, but work for their own profit, and at the bidding of "governments and capitalists." Tolstoy divided all the great nineteenth-century inventions into two categories. The first were things directly harmful to human beings and an affront to true Christianity, such as Gatling guns and torpedoes. The rest, including electric light, were harmless but "useless (and) quite inaccessible" to most people, a vanity of the rich. The reckless pursuit of science and material wealth, driven by selfish ego rather than love of God and one's fellows, had turned the "men of our contemporary European society" into "beasts, who fly all over the world on railways, and exhibit to the whole world, by the electric light, their brutish condition."[25]

Many who admired Tolstoy's sincerity thought his jeremiads against material progress were "undermined by an admixture of error, unreasonable absolutism, and extreme impracticability." In the United States, at least, the suggestion that electric light was a frivolous plaything of the rich, unavailable to everyone else, bore little relationship to reality. Many worried about the growing gap between the rich and the poor brought about by corporate capitalism. Yet the benefits of electrical power seemed widely democratized. By the early twentieth century, all American town dwellers could enjoy some of the pleasure and convenience of an electrified nightlife and a brighter workplace, while domestic lighting was coming within reach of many middle-class consumers and a growing number of urban workers. The wealthy had no motive to hoard electric light as a privilege of their class—the very nature of the utility business required a profitable company to achieve an economy of scale that lowered the cost so that many could enjoy, and pay for, electricity's benefits. As more central stations went online each month and the cost of electricity continued to drop, electric light seemed sure to become available to most homes in the more densely populated areas.[26]

In this respect, what distinguished the late-nineteenth-century technological revolution was not its creation of vast private wealth but the remarkable way its benefits extended to so many citizens. The modern industrial system built enormous fortunes for some but also served a more democratic purpose, improving "the average of human happiness" by providing mundane comforts to the multitude. Better educated, more inventive and self-reliant than many of their European counterparts, American workers at the end of the nineteenth century considered themselves entitled to some share in the bounty of the nation's expanding consumer economy. "What were the enjoyments of the rich have become the comforts of the poor," as one scholar explained it. "The seemingly absolute needs of a comfortable life have grown wider and deeper, until in the present day the day laborer finds himself, in his thirst for higher conditions, but vaguely content with conveniences many of which were unattainable and even undreamed-of luxuries to the rich and noble of a few centuries past." An economist who surveyed the spread of utilities into small-town America in 1891 concluded that "the homes of myriads of working people, mechanics and those in humble life" enjoyed hot water on tap, central heating, and gas or electric light, comforts not even the wealthy had enjoyed in the supposedly "good old times." "Had we been able to choose our lease of life on earth," he concluded, "could we have chosen a better date?"[27]

Though Americans felt lucky to be living atop the pinnacle of human evolution, most expected that much more would soon be accomplished as science achieved a greater mastery over the forces of nature and every old invention bred countless new ones. "An ever larger and larger number of fertile brains are continually at work in discovery and invention," as one engineer explained, "and these fresh brains start from an ever-widening vantage ground of accumulated research and proved experience. The result must surely be that important inventions and new discoveries will crowd thicker upon the world in the twentieth century than in the nineteenth." Edison agreed, predicting that humanity was "just at the beginning of inventions." And so it seemed incredible that anyone, even an ascetic Russian sage, could doubt that "today is better than yesterday," and that "tomorrow will be better than today."[28]

Even as the dynamo and electric motor transformed so many aspects of economic production, a surprising number assumed that this was just a trolley stop on the rails of progress. When humans could conjure so much unexpected power out of the magnetic field of a dynamo, it seemed likely that entirely new sources of energy would soon be discovered. "The forces of nature are set to do our bidding," as one engineer put it. "In the last one hundred years man's productive capacity has probably advanced more than in all the preceding years that he had inhabited this planet, and the revolution wrought by the development of the capacity to manufacture power has just begun." Anticipating the "exhaustion of the coal fields," some predicted that the future belonged to solar power, and experimented with the use of selenium cells. "If we can only devise means for imprisoning the heat of the sun," one American engineer wrote in 1887, "then we can abandon our coal mines and forget they ever existed."[29]

Many agreed that the steam-powered dynamo and incandescent light were too inefficient. As Edison put it, burning coal in a steam engine in order to produce electricity was an "expensive and vast waste" that had to be replaced by some of nature's "hidden powers still to be revealed." He predicted that the first one smart enough to solve the problem would be hailed as "the king of inventors," and he searched for ways to extract electrical energy from coal without combustion. Charles Brush experimented with wind power, erecting a massive windmill in his backyard that provided power and light for his Cleveland mansion. Some predicted that scientists would soon learn to harness electricity from the earth's magnetic field. Others forecast an energy revolution when chemists figured out what "the firefly knows," mastering the properties of phosphorescence and bathing rooms with a form of "luminous ether." Perhaps the most visionary of these futurists proposed that the entire planet might be "belted" with giant mirrors capable of gathering sunlight and distributing it across the dark half of the globe, using a series of "great optical pipelines." Satirists enjoyed lampooning these electrical dreamers, offering their own schemes to light the world by harvesting static electricity from the fur of fighting cats.[30]

A reading public hungry for news about the future provided a ready market for Victorian writers of science fiction and utopian literature, and many of their works envisioned what a world shaped by perpetual invention and inexhaustible energy might be like. These authors disagreed about whether humanity's technological tomorrow would be a lovely dream or a terrible nightmare, but none doubted that the future would be saturated with artificial light. One novelist envisioned the streets of a future New York "bathed in floods of soft, effulgent electric light." Abundant light at the flick of a switch marked the end point of an ancient human quest, though by the early twentieth century it seemed like a prophecy that required no special insight. For these Victorian visionaries, the brilliant light of the future served as a metaphor for a world liberated from want by the progressive impulses of science, producing a society that was more rational, orderly, and just. Light managed to convey these noble aspirations while also offering readers tantalizing visions of a world awash in the pleasures of a consumer economy. As one smug messenger from the future put it in one of these utopian novels, "We look upon the products of industry in the same light as one might who had a crop of apples falling off his trees." Another author captured this ideal more concisely in his title—"The World a Department Store."[31]

Those less sanguine about the direction of corporate capitalism never doubted that the future would be physically brilliant and overflowing with other electrical marvels, but predicted that it would come at the price of moral darkness. In his bestselling dystopian novel *Caesar's Column*, the Populist leader Ignatius Donnelly conjured an image of New York in the year 1988. Approaching the "mighty city" in an airship, the book's narrator finds a city of ten million, brilliantly lit by "magnetic lights" powered by electricity harvested from the aurora borealis. "Night and day are all one," the hero marvels, "for the magnetic light increases as the daylight wanes; and the business parts of the city swarm as much at midnight as at high noon." Initially impressed by the sight, he soon learns that such technological power comes with a steep political price, as the city is ruled by a rapacious oligarchy that, by the end of the novel, comes to a bad end. A civilization brilliantly lit by electric power ultimately burns in the flames of revolutionary violence.

Unlike Donnelly, most prognosticators assumed that more light meant a better life, a higher level of civilization. But in the early twentieth century, as Americans continued to wire themselves for an ever brighter future, a growing number of critics challenged this assumption. While most expected the electrified future to be more enlightened, orderly, and liberating, these others warned that America was creating a consumer culture of brilliant superficiality—not progress, but a descent into the bondage of a modern mass culture of stupefying glare.[32]

Night view of lower Manhattan, circa 1920.

Ten

Exuberance and Order

By the early twentieth century, the illuminated skyline of major American cities had become an unplanned work of art, splashes of light and color that coalesced into a sight widely hailed as the quintessential symbol of the modern age. As the urban planner Charles Mulford Robinson put it, "There is hardly a lovelier picture on earth than the night view of a great city—its thousands of lights twinkling in a mighty constellation. . . . Its stars sing together, and their song is of the might of ourselves."[1]

And yet observers like Robinson faced this paradox: American cities assumed this grand visual quality only at night, and from a distance, preferably some high point where the cacophony of signs and streetlamps blended into one vast but remote "constellation." As a reformer working to impose visual harmony on America's urban landscape, Robinson saw as clearly as anyone that, up close, the electric city's grand beauty faded.

Others also noted the technology's strange bargain, the trade of day for night. Surveying a New York building illuminated by a new set of lamps, a journalist complained, "One would think that the day did not exist and that night alone had occupied the attention of inventors. A structure which nowadays looks particularly fine in the evening makes a sorry display in the daylight, when ten times as many people are looking at it." Surveying the iron scaffolding of an "electric sign monstrosity," *Scientific American* concluded that "it would be difficult to conceive an object more vulgarly obtrusive and more exasperatingly ugly." However magical the lights might look at night, their spell was clearly broken once the sun came up, even on New York's famous Broadway. The writer Simeon Strunsky described the problem most poetically. "O Gay White Way," he exclaimed, "you are far from gay in the fast fading light, before the magic hand of Edison wipes the wrinkles from your face and galvanizes you into hectic vitality." The electrified city delighted the eye by night, but was a depressing eyesore in the cold light of day.[2]

A young generation of American painters disagreed, finding new aesthetic possibilities in the city's scruffy streets at every time of day. But they especially loved the look of its late hours, and embraced the challenge of capturing urban America's luminous nightscape. John Sloan, George Bellows and other members of New York's "Ashcan School" painted streets smeared by the flickering yellow light of a passing elevated train, the lure of gleaming shop windows, working-class crowds gathered under streetlamps, night workers under the arc lights excavating for the new Pennsylvania Station, and the shimmering jumble of carnival midways and Coney Island nights. Searching for a visual vocabulary that captured the charged sensory experience of modern urban life, these artists loved the intensified and unnatural colors of electric light, the simplified patterns of stark light and deep shadow, the impressionistic glimpses of city life as people moved in and out of the surrounding dark.[3]

Young urban realists felt energized by their city's electric landscape, but others with a more traditional artistic temperament saw a city losing its soul, "lost beyond redemption." As one social critic put it, "A trip through the weird canyon of Broadway after dark leaves upon the mind the impression of having spent untold ages in a limitless journey through some horrible depths of Hades, where the soul has been wearied and the eyes dimmed to semi-blindness by the insistent protrusion upon the mind of glare and ugliness."[4]

The first great public battle over the intrusion of electric light on the American landscape occurred far from the urban centers, in a crusade led by preservationists to rescue Niagara Falls from commercial degradation. Decades before the invention of electric light, the grand scenery of the falls drew visitors who in turn drew entrepreneurs eager to make a buck. Well before the Civil War, tourists found the place crowded with hotels, souvenir stands, dance halls, and so many billboards that the falls themselves were hidden from view—except for those willing to pay for the privilege.[5]

By the early 1880s, conservationists reserved a particular disgust for the "colored light nuisance," the very popular nightly spectacle produced when tinted spotlights illuminated the falls. On hot summer evenings ticketholders assembled at Prospect Point to watch the falls splashed with rainbow colors. In an era fascinated by the play of electric light on falling water, here was the grandest fountain spectacle of them all. While tourists evidently loved this "magical effect," landscape reformers like Jonathan Baxter Harrison found it "debasing, vulgarizing, and horrible in the extreme." The scarlet beam made the falls look like the discharge of a sugar beet factory, he fumed, while under a yellow light the great cataract reminded him of the bilge from a whiskey distillery. Some called the illumination "artistic," but Robinson dismissed this "rubbish" as the delusion of people who had lost the ability to appreciate God's creation and had replaced it with the modern world's "morbid and diseased pleasure." Worse still, parents brought their children to see the sight, poisoning their sense of beauty. "That young people and children should be exposed to the influence of such a spectacle," he thought, "is a matter for deepest regret and sadness."[6]

The move to rescue the falls gained momentum when hundreds of American and English luminaries joined forces to "Free Niagara," pressuring New York to create a state reservation in 1885. The park's first commissioner banned the evening light shows and supervised the construction of a landscaped public park designed by Frederick Law Olmsted, the architect of Central Park. This less commercialized encounter with the falls proved more popular than ever with visitors, while one guidebook boasted that the great landmark no longer suffered from "the dominant materialism of the age." For decades the light shows were only allowed for temporary exhibitions on special occasions.[7]

Even this modest victory over the spread of electric-powered commercialism was inconceivable in the downtown centers of major American cities, where companies spared little expense in the scramble to be noticed by potential customers. Electric signs had been around almost as long as the incandescent bulb itself; the earliest was probably the Edison Company's massive "Column of Light," featured in early electrical expositions, that flashed "EDISON," promoting both the brand and the man. Taking advantage of brighter and more efficient bulbs developed in the early twentieth century, sign makers introduced ever more elaborate innovations—new combinations of colored bulbs, the scrolling script of "talking" signs, and complex circuits that created the illusion of motion. The latest examples of this so-called spectacular lighting always drew a crowd. Even visually jaded city dwellers could not resist the incandescent illusion of racing electric snakes, bursting skyrockets, effervescing bottles of ginger ale, and Civil War soldiers silently playing the bugle. A New York cleanser company entranced pedestrians with an

Times Square, 1923.

....

ink blot of purple bulbs spread across the night sky, which was magically wiped clean by a swipe of red light, "the great spot appearing and disappearing . . . in that manner continuously." Sign inventors developed revolving drums to automate these circuits, but in the early years many were operated by hand.[8]

Some signs became major tourist attractions. The side of one New York hotel, for example, depicted a chariot race, the horses and riders twenty feet high in twenty thousand bulbs, linked by circuitry complex enough to leave New Yorkers "agog," marveling at the sign's animation of lifelike detail. The horses' legs churned "naturally," chariot spokes whirred, and the driver's crimson robe appeared to flap in the wind while animated clouds of incandescent dust trailed behind. Advertisers doubled their exposure when newspapers reported on the latest innovation in the sign maker's art, encouraging readers to head downtown to see "That Big Electric Sign" for themselves.[9]

Businesses were not the only ones hoping to use electricity to draw attention. Many cities and towns erected massive electric signs to welcome visitors and lure investors. Denver, for example, gathered public donations to erect a gigantic "welcome arch" at its train depot, greeting arrivals with the glow of two thousand colored bulbs. Civic and commercial groups commissioned signs to add a festive air to special occasions, welcoming returning war heroes, marking a dignitary's visit, or creating a celebratory mood at a convention of Elks or cyclists. Sign makers learned to "cash in on the spirit of patriotism" by helping civic groups erect massive electric flags, a tool to draw crowds on national holidays and to stir their sense of national pride through an awesome display of rippling red, white, and blue bulbs. Electric advertisers also found a brisk business serving churches looking to improve their Sunday market share with electrified crosses and illuminated signs; many of these used Gothic or Old English letters in an attempt to match their architecture and preserve a semblance of "dignity."[10]

At a time when the modern advertising industry was just in its infancy, electric signs epitomized the new consensus that the best way to reach a mass market was the constant repetition of colorful pictures. As department store magnate John Wanamaker put it, "pictures are the lesson books of the uneducated," and those that featured color and motion offered the most "eye appeal." While a new army of admen and designers applied this strategy to brand labels, magazine ads, and billboards, electric signs proved to be the quintessential "picture medium," the first form of electronic broadcasting, creating colorful moving images that gave companies a distinctively modern power of "compulsory attraction."[11]

While pictorial effects drew most attention, the signs also turned the night sky into a velvet page, imprinted with the frenetic and stunted prose of the modern sales pitch. Printed billboards and posters had done the same in American cities for many years, but everyone recognized that the surest way to "burn" a brand name into the public mind was to electrify it. Novelists in the early twentieth century tried to capture what was then a strange new sensation of commercial media saturation, describing sidewalk crowds hectored and hypnotized by the "flash-in and flash-out electric signs that kept breakfast foods and ales, the safety razors, soaps, and soups incessantly in the minds of a fickle public." New York's waterfront shimmered with signs big enough to read from New Jersey, "shouting aloud the worth of somebody's pickles or whisky." Companies invited their customers to participate in the process, holding slogan contests that gave lucky winners the chance to see their catchy propaganda emblazoned on the city skyline. Each letter cost money to build and more to power each night, so small businesses kept their message simple and direct; a sign that said "RESTAURANT" was an extravagance when one could get the same message across with the blunt command to "EAT."[12]

The enthusiasm for electric advertising soon spread to smaller cities and towns across the country, where residents hailed the arrival of each sign as proof of their hometown's growing sophistication. Often a sign's debut earned glowing press coverage. "The most admired thing in Butte [Montana] last night," as the local paper

Savvy church leaders installed lighting systems to lure customers and enhance the experience of worship. As one lighting specialist put it, "From every standpoint electricity is surely just as important in the church as it is in the place of amusement."

....

reported, "was the beautiful new electric sign. . . . Many declared it the handsomest sign in the city, and the sign blushed." When Springfield, Illinois, inaugurated a new sign for the *Chicago Tribune*, the paper boasted that this "wonder among signs" was bigger than anything in Chicago itself. Each night crowds gathered under the gleam of its two-foot-tall letters and enjoyed telling stories about how its powerful glow had unsettled their more rustic neighbors. On the first night it was lit, residents in surrounding towns telephoned Springfield to see if the city was on fire, while others believed they were enjoying a rare glimpse of the aurora borealis. A generation earlier, crowds had gathered to celebrate the arrival of their town's first string of arc lights. Now each of these massive new signs re-

The Edison Electric Illuminating Company of Brooklyn, selling signs.

newed the public's sense of wonder and excitement about electric light, and assured local boosters that they remained "among the first ranks."[13]

Leaving no space unmolested, inventors developed arc-light projectors powerful enough to cast words and images onto the sky itself. The Brush Company experimented with electric skywriting at the Columbian Exposition, projecting the daily attendance figures on the clouds each night, along with images of great men such as Christopher Columbus and Grover Cleveland. After the fair, Joseph Pulitzer set this spotlight on top of his *New York World* headquarters. On clear nights, a journalist reported, the company would compensate by "blowing steam into the air or sending smoke-producing rockets aloft." A New York department store installed a spotlight strong enough to be seen seventy-five miles away and threw "monster announcements in every direction." Surveying the spread of electric signs all the way to the heavens, an electric industry journal asked, with burbling enthusiasm, "has not advertising now reached its limit?"[14]

For social reformers working to create the "City Beautiful," electric signs had long since surpassed the limits of public decency and taste, spelling out America's crass materialism in vulgar slang that vandalized the night. To these men and women, every electric sign flashed the same message, that rampant individualism, self-interest, and lowbrow taste were running roughshod over the public's right to live in a world shaped by higher community standards of beauty and order. The novelist William Dean Howells complained in 1896 that the signs were defacing American cities as they tried to "shout and shriek each other down." What bothered him more was the fact that "no one finds it offensive, or at least no one says it is offensive." The critic Waldo Frank agreed. The brilliant lights of a street such as Broadway illuminated nothing, he wrote, but only blinded Americans to the glittering superficiality and shoddiness of their culture. The electric light is *false*, he concluded, "in the sense that it deforms."[15]

To these sensitive souls, the fact that so many Americans admired this "electric scream" raised troubling questions about their intelligence and their inner life. Would these people feel lonesome, one critic wondered, if their city buildings and empty suburban lots were not covered with "gigantic letters and lurid pictures?" Would they feel bored riding the train if their attentions were not constantly tickled by the "weird figures" on billboards and electric signs that lined every railroad track? These landscape reformers wondered how their fellow Americans could be so simpleminded and easily amused, cheerfully attending to signs that imposed "dumb recitals of indubitable values." Another critic compared the "jiggling abominations" of electric signs to a wedding present—"large, ugly, clumsy, heavy, expensive, and useless."[16]

Electric companies saw the matter quite differently, of course, arguing that their signs were not a needless blight on the landscape but agents of progress and prosperity. Selling signs and the electricity to power all those bulbs had become a lucrative business for the power industry, as merchants competing for the attention of urban crowds felt compelled to invest in ever more elaborate illumination. "In show window lighting and signs, electricity stands alone," one salesman boasted. "Storekeepers must have electric light for these purposes at any cost," even if that meant burning thousands of bulbs every night. Through signs and store windows, the most up-to-date merchants were learning to turn their own buildings into blazing advertisements for their business.

While companies put up these signs to command the attention of consumers, the cumulative effect served a civic function as well, creating the "bright and lively" streets of the modern commercial district, a social space where pedestrians felt safe after dark. More than public lighting projects, this unfettered electric advertising made American cities the brightest in the world, alive with a carnival playfulness of color and motion. Those in the sign business felt they deserved the country's thanks for providing the public with so much light, a catalyst for urban prosperity that cost taxpayers nothing. A city glittering with electric signs, as one explained, "receives the admiration of the broadest-minded men of the world."[17]

In the early years many signs had been hastily erected, and during storms some crashed down onto the sidewalks in a sensational fashion. Even the electric companies agreed that cities had a right to impose tighter restrictions to prevent signs from endangering the public by launching like sails in a storm. Many also supported a ban on signs hanging over the public sidewalks. Firefighters and insurance companies considered them dangerous, while merchants complained that projecting signs gave their streets a claustrophobic feel and blocked the view of other signs. Cities passed laws limiting the size of electric signs and requiring them to be fastened securely across the face of buildings or firmly erected on rooftops high above street level.[18]

The landscape reformers in the City Beautiful movement did not want these signs more firmly mounted—they wanted them gone, or at least carefully contained. Some suggested that all decisions about electric signs should be turned over to a civic art commission, a body of respectable and public-minded citizens who would impose aesthetic order on the American cityscape and condemn any sign that they deemed "a menace to the sanity of the population." Short of that, reformers thought it might help to impose a uniform color scheme on a district's electric signs, and particularly favored the idea that only white light should be used, producing an effect they considered "austerely beautiful, and refined."[19]

An editor at the *New York Times* dismissed the sign critics in New York's Municipal Art Society as elitists and reactionaries. True, the mass of competing signs gave entire zones of New York an "utterly commercial" look, but that was a case of truth in advertising, since that's just what their city was—"utterly commercial." Others who resisted the City Beautiful agenda pointed out that the utility wires and electric signs hung over streets already choked with plenty of urban eyesores and skies marred by billowing plumes of chimney smoke. America's built landscape was mostly an "unspeakable conglomeration of heterogeneous incongruities" and no sign, however ugly, was likely to make it look even "less artistic." And besides, from a proper distance the "spectacular effect" of all that colored light was quite beautiful, especially when refracted through filthy air.[20]

But as the skyline grew ever more brilliant with each passing year, many more critics began to wonder aloud if things had gone too far. Signs covered entire buildings in prime locations and soared hundreds of feet overhead. Using lurid colors and ever more complicated strobe effects, these broadcast commercials no longer invited attention but *compelled* it, coercing the "unwilling and resentful eye." As the sociologist E. A. Ross complained, "why should a man be allowed to violently seize and wrench my attention every time I step out of doors, to flash his wares into my brain with a sign?" Ignoring any standard of architectural or grammatical integrity, America's commercial electric skyline marked a dangerous new level of "riotous individualism in construction and ornamentation" that made its cities "objects of scorn" to European visitors, who came from more dignified cities where "at least a little regard is paid to the harmonizing of building with building and to the production of coordinated architectural effects throughout whole streets and squares."

Many Europeans did in fact make a visit to Times Square an obligatory stop on any American tour, and for better or for worse concluded that New York's fantastic nightscape epitomized the nation's culture. Some saw in those massive signs a nation enthralled, even enslaved, by its self-promoting capitalists, mentally infantilized by a nocturnal fantasyland. Surveying the hectoring signs in Times Square, the English writer G. K. Chesterton famously quipped, "How beautiful it would be for someone who could not read." The German filmmaker Fritz Lang was more positive, marveling at New York's "abysses of light, of moving, twirling, circling light that is like a statement of happy life."[21]

Critics of the City Beautiful movement sometimes accused its proponents of being romantics who longed for a bygone age of cobblestone streets and dim gaslights. But this coalition of urban planners and landscape architects insisted that they were not just naysayers who resented the intrusion of modern technology. In fact, they dreamed of illuminating the city in their own way, with an orderly and aesthetically harmonious system of streetlamps—one essential part of their plan to revamp the urban core with wider boulevards, parks, and civic monuments. As a first step toward that goal, they worked to clear the air of forty years of haphazard wire stringing and the tangle of poles erected by competing electric, phone, and telegraph companies, hoping to replace them with uniform, artful, and efficient new streetlamps.[22]

Since they lacked any legislative power, American advocates of city planning tried to win converts by example. Groups such as the Art Society volunteered their services to the city, hoping to convince politicians and taxpayers that they need not sacrifice beauty in favor of utility. Self-proclaimed guardians of "taste and reflection," these men and women struggled mightily against what they called "official indifference and public ignorance more dense and widespread than we had conceived possible."[23]

As part of their program, New York's landscape reformers held a competition for the design of an "artistic electrolier" to be erected on a pedestrian island in the middle of a busy stretch of Fifth Avenue. Believing that utility poles should be objects of civic art, the Society's judges reviewed over fifty design proposals. In a ceremony punctuated by "enthusiastic handclapping," they unveiled the winner, a fluted bronze shaft capped by three animal heads, "a group of children gamboling in the nude," and five large incandescent globes. Members of the Art Society believed they had provided the city with a model for "future work," an unlikely dream since the winning lamppost had taken skilled craftsmen two months to build.[24]

A few years later a speeding fire truck wiped out the Art Society's model lamppost. In spite of this setback, City Beautiful reformers had reason to feel that their campaign was paying off. Across the country, evidence suggested that citizens were growing tired of the eyesores of modern urban life, while some more forward-thinking merchants were beginning to recognize the power of good lighting and pleasing design to draw customers and raise property values. In Los Angeles, for example, the Broadway Improvement Association raised funds in 1905 to erect 135 of the more elegant lampposts, each mounting seven incandescent globes, an effect hailed by the *Los Angeles Times* as "a glimpse of dreamland." On opening night a vast crowd showed up to watch the mayor flip the ceremonial switch, and when the light came on cannons blasted, sirens sounded, and "huzzahs" rose from "thousands of throats." Within the year, the merchants on the city's other major boulevards organized to install their own new "white ways," eager to keep pace in the march toward both progress and profits.[25]

Like Los Angeles, many other cities paid for "white way" lighting through a subscription; an "improvement association" gathered funds from local businesses while the city agreed to pay for the higher electricity costs once the new poles had been erected. The improvement association in Chattanooga twisted the arm of any merchant who refused to pitch in by mounting the new ornamental poles in front of their property but refusing to turn them on,

The Municipal Art Society's award-winning streetlamp.

making it obvious who had been too stingy to help fund the city's new lights.[26]

The movement drew further energy from a generation of young Americans who studied urban planning and design in Europe, particularly in German universities, and brought home new ideas for imposing a visual order on American cities. In Germany it was "taken as a matter of course," as one of these American students reported, "that public and private work shall not be ugly." Part of a wider progressive movement to reform American cities, these young

A "white way" in Los Angeles, 1912.

idealists returned from Europe convinced that the American landscape need not be "glaring and crude."[27]

And so even as the electric sign craze spread across the country in the early twentieth century, a coalition of progressive reformers, conservative aesthetes, and enlightened merchants worked to impose a different language of light on their cities, investing in civic lighting systems that featured buried wires, sculpted streetlamps, and public parks and civic buildings washed in strong white light. In hundreds of American cities, boosters pointed with pride to their town's new "white way." The rows of bright lamps neatly lining their Main Street served as a visual symbol of the town's progressive spirit, its participation in the era's embrace of urban improvement. In their eagerness to electrify at any cost, towns had once accepted an ugly mesh of wires, haphazard wooden poles, artless lamps, and the brutal cutting of urban trees—in those early years, all that had seemed like the inevitable price of progress. But now a new generation of forward-thinking merchants and civic leaders seized the moment to recast the main boulevards of their downtowns, installing uniform street lighting that was not only functional but appealed to "the senses of beauty, harmony, and art." Larger cities invested heavily, holding design competitions for new lampposts and erecting miles of them on major boulevards. Eager not to be left behind, small towns scraped together enough funds to install at least a few dozen of the more elegant standards along their main streets, enough to proudly declare themselves a "white way" city. The thickening web of overhead wires had once been considered a sign of progress but, as Charles Robinson declared in 1901, "A city is now held most progressive when it shows fewest wires, not when it presents their greatest network."[28]

And so through the early twentieth century, as the sun went down and the lights came on, Americans saw two competing visions of the good life, a conversation in light between rival conceptions of democracy that had deep roots in the nation's culture. The exuberance of colored signs expressed the nation's free-market liberalism, its faith that the unrestrained pursuit of private gain would cohere into a shared culture that liberated individual initiative and creativity while serving the interests of a nation of empowered consumers. Skeptics have pointed out that this may be another of the electric signs' seductive illusions, as the nation's largest corporations quickly monopolized the skyline, owning the biggest signs and their prime locations—in places such as Times Square, at least, the forest of signs was a market, but not a free market. But for many the new signs brought a bit of Coney Island's democratic energy and fun into their daily lives, a technological exuberance that seemed distinctively modern and American.

Chicago's State Street, 1926.

Beneath this glittering, undulating mass of colored light stood the orderly rows of "white way" streetlamps, embodying the ideological counterpoint that historians have called civic republicanism. These more restrained white lights spoke not of private gain but of shared community purpose; they were illumination for citizens gathering in the public square, not consumers eager to be entertained or seduced into spending money. "Good street lighting creates a psychological impression of thrift and progress," as one advocate put it, "advancing civic pride, attracting favorable publicity, and promoting other improvements." Even the design of these streetlamps embraced the history of republicanism, the fluted pillars, decorative capitals, and symmetrical globes exemplifying the American Republic's fascination with the aesthetic values of Greek and Roman civic architecture.[29]

But in the early twentieth century, a new group of professional

By the 1890s, craftsmen offered wealthy customers a variety of artistic "electroliers."

lighting engineers offered a third alternative—not the gaudy glare of commercial light, or the staid classicism of the "white way," but an approach that they called "illumination." Like so many other aspects of social and economic life in the early twentieth century, lighting came under the purview of trained experts, new professionals who made a bid for authority in the marketplace based on their claim to bring expertise to bear on a pressing social need. Working to identify and then master the scientific and aesthetic principles of effective lighting, these specialists in "illumination engineering" devoted themselves to extending the reach of electric light into every corner of daily life.

Eleven

Illumination Science

Like other emerging professions in the early twentieth century, illumination engineering developed on multiple fronts. Enjoying a near monopoly in the incandescent lighting business, General Electric invested in research to make lamps more efficient and to open new markets; some universities added lighting design to their engineering programs, acknowledging the growing importance of artificial light in city planning and architecture; and in 1906 those who claimed some expertise in this new field organized themselves into the Illuminating Engineering Society. Modeled on professional organizations already established by civil and electrical engineers, the IES allowed members to network through their own journal and national conventions.[1]

For utility companies and electrical manufacturers, the move from selling lighting to selling illumination marked an important step in the evolution of their business—a new focus not just on the amount of light produced, but also on its usefulness and beauty. In the industry's early years, many customers made the big investment to have their streets, stores, or homes wired, only to find themselves dissatisfied with the results—the light was better, but not all that they had hoped. On major building projects, architects also expressed frustration. Too often, the addition of electric lighting undermined their work, its "lumps of light" dangling from fixtures that distracted the eye from the building's lines and decorations.[2]

Illumination engineers blamed this problem on the fact that most city planners, architects, and electricians did not understand the basic principles of lighting design, axioms that were only then being worked out. As one electrician facetiously remarked, 90 percent of lighting installations seemed designed to give "the least quantity of effective lighting for the money expended." In their product's first decades, utility companies had promoted it by unleashing an army of untrained "lamp peddlers," while builders and architects knew little about lighting and treated the matter as an afterthought. But in the first decade of the twentieth century, the lighting industry hoped to expand its market by offering customers the service of illumination specialists who had mastered the long neglected art of "placing lamps where they will do the most good."[3]

Among the leaders of this emerging field of illumination engineering, none matched the swashbuckling panache and righteous fervor of Laurent Godinez. Born in Poughkeepsie, New York, Godinez studied engineering, worked for the New York Edison Company, and then went into business for himself as a lighting consultant. For Godinez, better lighting was a *cause,* and he brought this message to the American public through a series of books, articles, and lectures, which he was prepared to deliver in English, French, or Spanish. Though his tours were often sponsored by local utility companies eager to juice the public's interest in buying more light, he always made clear that he was no "salesman of lamps," no shill for the electric industry. Rather, he often warned the public to be on guard against charlatans who betrayed the public trust by selling "worthless illuminants." This celebrated engineer attracted large and enthusiastic audiences for a two-hour technical demonstration of the latest insights of illumination engineering—a brand-new profession he considered among the modern world's "highest callings." Godinez was artificial light's angry prophet, as well as its poet. Too often, he thought, Americans had used the electric light to produce aesthetic abominations that damaged the eyes while blighting the landscape. But illuminators would change all that, showing how the technology could be used in the quest for "man's most perfect creative effort, the city beautiful."[4]

Godinez offered detailed suggestions on the proper illumination of everything from grand museums to humble kitchen pantries. He spent much time considering ways to improve the look of city streets, a project that he believed required a forward-thinking partnership between government and local merchants. Godinez welcomed the visual excitement of a well-designed electric sign, but loathed those "utilitarians" who slandered the lighting profession by throwing up shoddy work. He also preached against the timid traditionalism of the City Beautiful movement's popular "white way." He found its predictable rows of sculpted iron lampposts terribly inefficient, throwing more light skyward than on the sidewalk below. Further, Godinez complained that when every city and town used the same lighting scheme, the end result was "deadly monotony." Cities could not hope to create well-lit streets just by ordering a set of "ornamental columns and balls" from a lighting catalog. Electric light was no longer a novelty, he explained, and "the American public will not be attracted like a horde of insects by a vulgar glare."[5]

Instead, cities needed the services of illumination experts who could help them take advantage of all that the new art and science had to offer. Every street and storefront posed a unique set of lighting challenges, and only a professional knew how to use the latest technology to create city streets that were "expressive of individuality, and of permanent value to a community."[6]

From the Edison Company's earliest days, its lighting expert Luther Stieringer had used expositions to prove the technology's potential as an art form. The next generation of illumination engineers acknowledged him as a founding father of their cause. And yet decades after Stieringer's pioneering work, those seeking to ground the lighting profession in the authority of science felt humbled by all they had yet to learn about artificial light's effects. "The ascertained facts are few—all too few," as the head of England's illumination engineering society put it. "Their significance is immense; their economics and social value great; but the ignorance respecting them generally is colossal!"[7]

Those seeking a greater mastery over artificial light first needed better ways to measure it, a shared language that reflected more precise and objective standards. Lacking anything better, the first electricians used the unit of "candle-power," but candles varied and the tools for measuring the light were primitive. In the first decade of electric lighting, the claim that an incandescent bulb delivered the equivalent of sixteen candles seemed clear enough, but as artificial light grew ever more powerful and sophisticated, buyers and sellers needed something more precise. As one frustrated engineer summed up the problem: "To say a light is of a certain candle power, is about as scientific a description as it would be to characterize a cat as twice the size of a kitten."[8]

In search of a common nomenclature and illumination standards, lighting companies and their engineers held international meetings to craft a more subtle and mathematical vocabulary of light measurement, while research laboratories invented a range of photometric tools such as the "illuminominator," the "colorimeter," and the "lumichromoscope." In the early decades of the twentieth century, engineers produced increasingly sophisticated research on the properties of lighting, testing various systems in the laboratory and in real-world environments. Marking a paradigm shift in the field, these researchers focused not on light but on *illumination,* looking away from the lamp itself to study its three-dimensional reach and its effect on the objects around it. They placed bulbs in different locations in a room, calculating the light's ricochet, and developed mathematical formulas and graphs to capture the effects of distance, reflection, angle, and room color. While new measurement tools provided a more precise mathematical description of various lighting systems, the customer' s eye was the ultimate judge, and so researchers also explored the physiology of vision and the psychological impact of light. Studies measured the "eye fatigue" caused by various forms of flicker and shadow, the demands that various activities placed on the eyes, and the moods induced by different colors.[9]

The move to professionalize illumination came in the midst of a second revolution in the quality and quantity of electric light. Following the lead of German chemical engineers, General Electric's research laboratory introduced a tungsten-filament bulb in 1907 that was much brighter than carbon-filament lamps, and up to four times more efficient. Some utility owners feared that this improvement would soon destroy their business, undercutting the demand for electrical current. Other electricians correctly predicted that Americans would rather have more light than smaller bills, and they welcomed tungsten bulbs as the tool that would finally let them offer electric light at prices that could compete with gas. From a technical standpoint the new metallic filament made Edison's origi-

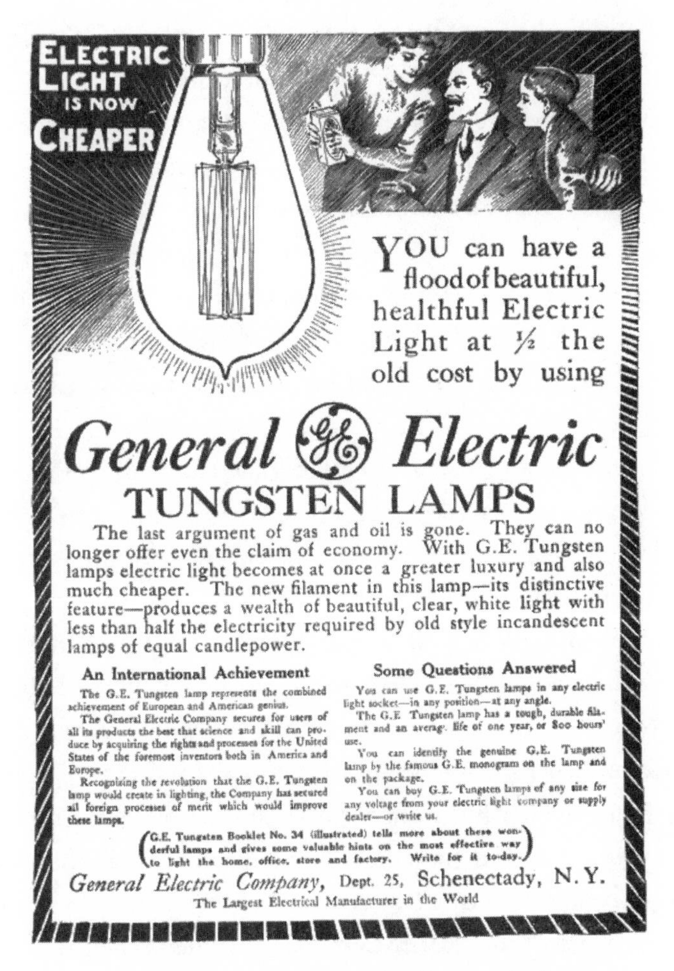

Marketing its first metallic-filament bulbs, General Electric announced that "Edison's Dream has come true"—cheap and abundant light for all.

....

nal carbon design obsolete, but the transition to the improved technology was gradual. Utility companies included free carbon bulbs as part of their electrical service, but customers who wanted the more expensive tungsten bulbs had to buy their own. This barrier slowed but never stopped the growing consumer demand for the latest technology, and the brighter world that it offered.[10]

As people switched to the new tungsten filament, brilliancy was no longer the goal but the problem. Any old-school electrician could now rig enough bulbs to flood a room with light. But the illuminators found that good vision required more; it depended on a proper contrast between light and shadow, a balance between diffused and direct light, and, most of all, attention to the very modern problem of "glare," the experience of pain or discomfort produced by *too much* light.

The demand for greater scientific understanding of the physiology of vision received a boost from a national health scare sparked by medical experts who blamed electric light for an alarming growth in "defective eyes." Americans might live in the brightest cities in the world, as one doctor put it, but they also suffered more "ocular exhaustion, discomfort and congestion." Cooler heads suggested that modern people's vision was no worse than that of any previous generation—they just lived in a world that demanded greater visual acuity. But from the earliest days of electricity some doctors warned that the new light "unduly stimulated" the retina, stretching nerves to a pitch of constant excitement. "No eye can sustain the glare of the electric light that now meets the gaze everywhere," as one complained in 1890. "It is ruinous to the eyesight, and induces paralysis to the eye." Even worse, he predicted that this plague of spectacle wearing would be passed down through the generations, producing over time a race of dim-sighted Americans. If these critics were right, more light was gradually producing less vision—an irony that struck one indignant engineer as an absolute "perversion of the logical order of scientific development." Such dire warnings that electric light was going to "wither and wreck lives innumerable" stirred ripples of anxiety in the culture.[11]

. . .

Lighting specialists denounced the "pernicious idea" that this new technology was causing the nation's alleged eye problems. Electricity's strong and unwavering light actually reduced eyestrain, they countered, and they blamed the misunderstanding on the public's confusion about how to use a light bulb. "The whole trouble with the electric lamp," as one explained, "arises from the manner in which it is used, rather than in the lamp itself." Too many customers judged the merits of electric light by staring straight at a naked bulb, instead of considering how well it illuminated the objects around it. Like domestic suns, the modern incandescent lamps were too strong to be stared at, their light best moderated through lampshades and reflectors.

Self-evident today, the proper use of an incandescent lamp is a social practice that, according to one electrician, was misunderstood by 99 percent of Americans in the early twentieth century. Why pay so much for electric light, these customers surely wondered, only to hide it behind a shade or to place it out of the line of sight, like the Bible's proverbial lamp under a bushel? Such an idea must have seemed like the scheme of unscrupulous electric-current salesmen eager to sell customers more light than they needed.[12]

And so the new illuminators faced not only technical challenges but an educational one, teaching the public how to use light well, preventing glare from overwhelming their eyes and defacing their culture. Members of this new profession shared a sense of mission, sure that their technical mastery of light would yield great public benefit. They aimed to "harness" the power of light in the same way that more traditional electrical engineers had channeled the energy of the Niagara River. At a time when other progressive reformers rallied to defend the nation's natural resources against the ravages of an unregulated free market, the illuminators aimed for nothing less than "the conservation of vision."[13]

Since the problem of glare could be traced most often to the eye's direct exposure to the strong light of the new tungsten filaments, illumination engineers agreed that as much as possible, "lighting should be from sources concealed." The self-proclaimed prophet in this new field of indirect lighting was a Chicago inventor and entrepreneur, Augustus Darwin Curtis, founder of the X-Ray Reflector Company. Declaring himself "the archenemy of glare," Curtis designed and manufactured silvered glass shades—he felt the proper and more positive word for it was "reflector"—that would shield the user from the bulb's "dazzling brilliancy" while concentrating the light on its proper target. Curtis showcased his invention in the store windows of the Marshall Field's department store in Chicago, and soon retailers across the country adopted his silvered reflectors. For interiors, Curtis and others designed fixtures that bounced light off the ceiling, luminous pendant bowls that sacrificed efficiency but produced a diffused light that was more comfortable to the eye. Commonplace today, these fixtures were welcomed by the lighting industry as a conceptual breakthrough that promised to finally deliver "lighting at its best."[14]

The same principle of indirect lighting served in one of the grandest lighting projects of that era, the illumination of the Woolworth Building on Manhattan's Broadway, the tallest occupied structure in the world when it opened in 1913. A dozen lighting experts and forty electricians worked for months, carefully hiding six hundred incandescent car headlamps on balconies and behind screens, more candlepower than most small cities used for their entire territory. The first generation of lighting designers and architects had studded the lines of grand buildings with strings of naked carbon-filament bulbs, but the Woolworth Building reflected a new approach made possible by powerful floodlights. At night, the tower

Woolworth Building, circa 1920.

glowed with an even coat of strong white light that drew the eye but never distracted it, amplifying the rich gilded details on its white terra-cotta surface. Hailed by one lighting expert as a "torch of civilization," the building culminated in a summit that pulsed red and white beams, glowing like a "scintillating jewel." The success of the Woolworth Building showed the value of close cooperation between architects and illumination engineers, proving beyond a doubt that imaginative lighting could make a grand building look even grander after dark.[15]

Illumination engineers still complained that many architects showed a "peculiar reluctance to be educated" about lighting design, many of them resentful of the way bad lighting marred the visual integrity of their buildings but unwilling to pay for better advice. At the same time, many old-timers in the lighting business scoffed at the idea that illumination required some specialized technical expertise, calling the upstarts "useless individuals" who were trying to cut in on their business, charging customers for common sense dressed up in meaningless jargon about "energy curves, luminosities and pupillary diameters." The illumination engineers countered that the old breed of electricians had succeeded all too well in their basic mission. True, they had created technical systems that produced ever more light, but since they lacked a basic understanding of lighting principles they had also produced the "most wastefully used and carelessly handled of all modern utilities." As one summed it up, modern Americans could now enjoy abundant light, "but it wants distributing."[16]

Long before illumination engineers laid claim to being artists of light, the first generation of designers had explored the aesthetic potential of incandescent lamps. From the start, Edison had worked with manufacturers to offer customers a wide range of decorative fixtures, and at early exhibitions his team had been quick to demonstrate some of the design advantages afforded by incandescent bulbs. Since they cast little heat and no exhaust, they could be entwined with flowers, shaded by paper lanterns, hung in strings and clusters, and even submerged in fishbowls. Still, most lamps mimicked the traditional forms of oil and gas lamps, leading the new generation of illuminators to complain about this lost opportunity to realize the light's decorative potential.[17]

By the early twentieth century, however, more artisans embraced the challenge, creating lamps that expressed the changing aesthetics of the modern age. Most agreed that making lamps beautiful meant hiding the bulb itself. Louis Tiffany and other practitioners of art nouveau produced popular lamps that filtered incandescent light through richly colored patterns of translucent glass, camouflaging the age's most sophisticated technology inside bouquets of lilies and morning glories. Outside on the street, electricity's delivery system scarred the landscape and defaced buildings and trees—but in the hands of Tiffany, the lamps themselves offered the compensation of a rare beauty.[18]

Illumination engineers welcomed these innovations, but they were less interested in making lamps into art than in finding ways to artfully arrange the light itself. Lamps were just building blocks, as they saw it. The real challenge involved selecting the right mix of bulbs and fixtures, creating the "moods and illusions" best suited to every room. As historian Chris Otter has pointed out, late-nineteenth-century city dwellers lived in an intensely visual environment that demanded a range of seeing skills; they had to identify and read faces in a crowd, decipher signs, read scientific instruments and newspapers, and navigate crowded streets and complex social situations. Illumination researchers studied the visual challenges of daily life in urban America, along with the more specialized demands of everything from bowling to blacksmithing, and published their recommendations for proper lighting intensity, color, and lamp distribution in their own professional journals. In the process, they invented a grammar of artificial light designed to shape and enhance every aspect of daily living—not just to capture attention, but to serve the eyes in a wide variety of tasks and settings.[19]

No human activity engaged their interest more than shopping. Savvy merchants had always used light and shade as tools of their trade, to reveal their best goods and to conceal the rest. Using early arc lights, Wanamaker had shown how effective brilliant light could be as a "creator of crowds." But illuminating engineers developed much more sophisticated lighting strategies. Studying the art and science of grabbing a customer's attention and provoking a sense of desire, they made electric light a crucial component of modern salesmanship.

Aided by the invention of steel-framed buildings and improvements in plate glass, urban retailers created ever larger shop windows and learned to use them as advertising spaces, taking advantage of the working public's "natural tendency to enjoy displays at night as a sort of relaxation." Before, owners had often filled their windows with ill-lit mounds of goods, a visual catalog of the store's merchandise. Some drew attention by ringing those windows with rows of naked bulbs; once an effective novelty, that approach struck the new generation of lighting experts as "wholly unscientific, not to say barbarous." Now, with the help of illuminators and professional "window trimmers," merchants experimented with new forms of display. They installed specially designed lighting fixtures and used theatrical techniques to showcase a few select items, luring customers with their windows' dramatic visual appeal. Like the massive electrical signs that hung above the streets, these colorful displays drew crowds each night.[20]

Laurent Godinez preached that the key to arresting the attention of sidewalk window shoppers was *novelty,* and he turned this

The storefront as electric sign, Washington, D.C., circa 1919–20.

commonsense truism into a scientific principle. Supported by the merchants of the Jersey City Chamber of Commerce, he made the shop windows on a downtown boulevard his laboratory, measuring the "attraction factor" of various lighting schemes. He rigged the shop windows with curtains, concentrated spotlights, and blended colors and flashers, then stepped back to measure how many pedestrians stopped, how many crossed the street to investigate and—the ultimate goal—how many entered the store itself. He was particularly proud of an experiment that proved he could draw crowds to a window that contained "nothing but *light*." He suspended a blank white board, illuminated by concealed spotlights, and "attracted so much attention that it nearly caused a riot." He estimated the "attraction factor" for this window, "displaying nothing," was 100 percent. The medium had become the message.[21]

Lighting inside the shop mattered too, not only helping to exhibit the goods but stimulating customers, energizing their desire. As shoppers grew more accustomed to strong light in public spaces, a store that failed to keep up looked dim, shoddy, and old-fashioned. But once again, illuminators warned shop owners against using too much light. In the early years, a naked arc lamp could draw customers who enjoyed the novelty. A generation later, lighting experts advised that harsh white light made goods look cheap—especially if they *were* cheap—while the glare from exposed fixtures could drive away customers.

In larger stores, new systems of indirect lighting provided abundant but diffused light that soothed the eye, making the showroom more inviting. But illumination experts refined this idea with special lighting for each kind of goods, a strategy to draw customers' attention and stir their longing to buy. Jewelry and cut glass sparkled best under clean white spotlights; a diffuse white light worked best on silver, guns, furs, and lingerie; and a warm yellow light was recommended for everything from candy to cast iron, shoes, stoves, and furniture. Grocers also learned which tints made their produce look freshest and their meat less cold and dead. The wide use of illuminated glass showcases, enticing customers to see without touching the merchandise, also created new opportunities for the lighting designer's skill. Then and ever since, modern shopping often meant gazing at artfully lighted objects arranged in glowing boxes of glass.[22]

While lighting experts denounced what Godinez called urban America's "crass commercialism," most were enthusiastic supporters of the nation's emerging consumer culture. They aimed not to challenge the public's growing hunger for all the goods delivered by corporate capitalism but rather to improve the experience, sure that their art could stimulate a love of beauty that would enrich the customers' lives while ringing up profits for their merchant clients.

Lighting specialists also played a role in refining the production process, which called not for seductive illusions but for increased precision. Studies found that the majority of factory buildings were badly lit, a problem that produced a quarter of industrial accidents, damaged workers' eyes, and caused mistakes that wasted millions of dollars each year in spoiled materials. While many factories continued to depend on natural light, the sun's rays proved too unreliable for all those industries that demanded greater efficiency. Illumination engineers sold their services to manufacturers by insisting that good artificial lighting was "an investment, not an expenditure," one that would pay for itself by reducing waste and preventing accidents. Workers moved more slowly and cautiously in dim rooms, while lazy employees found it easier to shirk, even nap, in dark corners. Once again, illuminators insisted that the solution was not just more light but a *smarter* light, engineered for the specific demands of each workplace.[23]

More controversial was the illuminators' claim that better lighting had a "stimulative" effect that improved workers' attitudes toward their jobs, "and even toward life itself," an emotional adjustment that made them more productive, less burdened by fatigue and depression, and "more loyal to the interests of the firm." While General Electric's experts assured factory owners that improved scientific lighting would produce a "stupendous" jump in worker efficiency, a hotly debated 1927 study could find no conclusive evidence to support that claim.[24]

In these same years, the demand for night work continued to expand in many industries, a trend that accelerated when America entered World War I, creating urgent demands that led a number of factories for the first time to add a second and third shift. Studies showed that accident rates doubled during the night shift, with one concluding that one hundred thousand workers were "incapacitated" every year due to accidents linked to inadequate lighting. As part of a wider movement to pass new worker-protection laws during and after the war, the industrial commissions in many states drew on the studies done by illumination engineers, incorporating these findings into lighting codes, for the first time treating access to sufficient light as a fundamental safety concern for industrial workers.[25]

The nation's public schools also proved fertile ground for lighting reform. In 1914, Laurent Godinez concluded a five-year study of school lighting by identifying no fewer than *four thousand* different problems in these "bad eye factories." The doctor and educator F. Park Lewis was among those experts who warned that forcing children to study in bad light was rapidly turning the country into a "bespectacled nation." A growing number of students attended high school in the early twentieth century, using a curriculum that required hours of reading and other "close work" each day, stressing their eyes at a vulnerable point in their development. "We have made a fetish of books," Lewis warned, becoming "a reading rather than a thinking people." Along with other educational reformers, he urged schools to offer more manual and physical education, and to invest in better lighting. As part of a wider program to improve "educational hygiene," illumination engineers joined forces with architects, the American Medical Association, and state legislatures to issue new guidelines for classroom lighting. These included detailed recommendations for bulb size, fixture placement, and the proper illumination of chalkboards. As part of their war on *glare,* that "omnipresent bugbear of school illumination," they even pressed publishers to stop using glossy paper in their textbooks, which bounced strong light into defenseless young eyeballs.[26]

While schools required clear, diffused light, religious sanctuaries posed more complex challenges for illuminators. Seeking an audience in a diverse and competitive religious marketplace, many early-twentieth-century church leaders found it prudent to consider not only their parishioners' spiritual needs but their physical comfort as well; they installed lighting systems designed to enhance feelings of reverence, while also creating a mood that was "inviting, cheerful, and comfortable." In church no less than in the theater, modern audiences demanded more light, but Godinez complained that too many churches went for the "typical gin-mill-blaze-of-glory effect," drenching their pews in a glare that ruined architectural effects and even lulled parishioners to sleep during the sermon. Illumination experts worked to retrofit venerable old sanctuaries, hiding the lamps entirely or integrating them into architecturally appropriate fixtures. As A. D. Curtis put it, exposing an ugly light bulb in church was as inappropriate as "a harsh word loudly spoken."[27]

Different denominations handled electrification quite differently. Lighting designers found that evangelical churches preferred a strong, bright light, one that felt practical, inclusive, and symbolic of the centrality of the scriptures. Those sects that relied more on liturgical ritual used light more sparingly and symbolically, preserving a bit of what Godinez called the ancient "mysticism of gloom." American Catholics followed the dictates of the Vatican, which temporarily banned the use of electric light in churches in 1889, finding it both dangerous and theologically suspect. Several years later, however, some of New York's Catholic churches showcased the new light's symbolic possibilities, inspiring their parishioners with floodlit pulpits, illuminated statuary, and angels holding incandescent "sacred hearts" while wearing diadems of electric bulbs. Still, Catholic Church doctrine stressed that electric light had no place on the altar, where wax candles remained orthodox. As one Catholic leader explained, "A display of lights, artificially arranged so as to attract attention to itself rather than to the center of worship, which is the Real Presence, would be an abuse. People would say 'Look at the lights on the tabernacle!' instead of 'Look at God in the humble Host!' "[28]

After much debate and soul-searching, orthodox Jews concluded that turning on an incandescent lamp violated the sacred injunction against lighting a fire on the Sabbath. Mormons, by contrast, were early and enthusiastic adopters, among the first customers for Brush's arc-lighting systems. When they erected a gleaming white granite temple in the center of Salt Lake City in 1892, it included its own power station lighting five massive incandescent electric chandeliers. A thirteen-foot gilded statue of the Angel Moroni, a lamp in his crown, topped the building's highest peak, spotlighted to make it the most visible point for miles around. Visitors to the Mormon community often commented on its enthusiastic embrace of the latest technology, which even some skeptics took as a sign of Mormons' "unbounded faith in themselves and their future."[29]

While lighting designers experimented with a large and expanding range of lighting arrangements, searching for the best combinations of color, shade, and light intensity for each human activity, GE's lighting expert Matthew Luckiesh felt that the whole matter boiled down to a distinction between white and yellow light. Throughout the long arc of evolutionary history, he argued, humans had worked by the strong light of day; in the modern workplace, then, artificial light should strive to mimic the clear white light of the sun. After work, however, humans had always spent their leisure hours in the more comforting yellow light of a flickering flame. At this late date, Luckiesh thought, no modern lighting designer could much change a relationship to light that had been hardwired into human heredity, an association between warm yellow light and the "esthetic" dimension of the human soul. This was the light of art, of meditation and contemplation, of fun and companionship, and of domestic bliss.

Historians examining Victorian ideas about domestic life suggest that those urban dwellers who could afford to often constructed their homes to be havens, woman-guarded sanctuaries from an urban street life that seemed each year to grow more crowded, polluted, and threatening. The working classes, jammed into small dark rooms, threw open doors and windows, conducting much of the business of life outside, in streets and parks and glittering saloons. But middle-class Victorians retreated to their interiors, seeking to create "soothing environments and private worlds." As the novelist Edith Wharton put it, in the modern city privacy had become "one of the first requisites for civilized life."

In search of that elusive privacy, homeowners often draped their windows with heavy velvet curtains, blocking out the street noise but also much of the sun. Strong interior lighting became more important than ever. Just as the house became sharply divided from the public life bustling around it, the rooms within carved up zones of public and private life—the parlor for company, the nursery for children, a den or library for the man of the house, the bedrooms and dressing rooms for private and intimate life. In American homes, boarders and extended families became less common and servants moved to more distant quarters of the house, often reached by a narrow back staircase.[30]

This trend had begun well before the arrival of electric light, but the new technology accelerated the development of the modern, private middle-class household. Before gaslight began to enter urban homes in the mid-nineteenth century, only the wealthy could enjoy strong light in their homes on a daily basis, and marked special occasions with an extravagant blaze of spermaceti. The middle classes burned candles and oil lamps much more sparingly, taking their evening meal in the late afternoon and gathering at night for reading and sewing around a central table lamp. Gas made strong light available for the first time to middle-class homes, and many embraced this not only as a household convenience but as a status symbol. More of the house became usable after dark, and the cost of evening entertaining went down.[31]

In the early twentieth century, the spread of electric light and improvements in gaslight produced a steady drop in the annual cost of lighting, even as middle-class households enjoyed more light than ever. Here in the domestic sphere of a well-appointed home, the new breed of lighting designers found much work to do, offering guidance on the proper way to illuminate every room, from the front porch to the maid's quarters, from the coal bin and fruit cellar to the attic stairs. They urged homeowners to consider the value of installing ambient lamps to fill an entire room with soft light, and more focused lamps that served the greater visual demands of reading, sewing, and grooming. The piano called for its own specially designed lamp, and light salesmen touted the value of an electric lamp beside the bed, which offered the first responsible way to read late at night without the risk of burning the house down.[32]

Again, the untrained and heedless current merchants of electric lighting's early years had done much to damage the industry's reputation, at least among more sensitive and demanding consumers such as Edith Wharton. As late as 1897, she advised her readers that a well-appointed home should avoid electric light altogether, since it offered all the charm of a train station. On the cusp of a radical change in what Americans would consider sufficient light, some still spent their evenings in a drawing room lit only by a single lamp, preferring its "coziness and tranquility." But most who installed electricity never looked back, welcoming the chance to fill their homes each night with enough light to allow the use of entire rooms, free of any candle and unencumbered by the dark.

Improved wiring schemes were equally important in transforming the use of light in the home. With much cheerful guidance from the utility companies, customers learned the value of installing a switch at the entrance to every room, allowing one to move through the house and "never, for a single second, be in the dark." Press a button, the ads breathlessly informed potential customers, and "presto, the lights are turned on. . . . Isn't this simple and wonderful?"

Electric light enabled people to move more freely in their homes after dark, and thanks to the early-twentieth-century invention of wall plugs, the lamps themselves became more portable. The first generation of lighting fixtures had been as firmly mounted in each room as the old gas fixtures; homeowners could use portable appliances and lamps only by screwing a flexible cord directly into one of the hardwired wall or ceiling lamp fixtures, an arrangement that produced what one historian has called "an almost carnival festooning of cords in the home." But as the system evolved in the early twentieth century, inventors solved that problem by developing a variety of more convenient wall receptacles, the ancestors of today's standardized outlets and plugs, an element in the modern system largely unanticipated by Edison and his first-generation rivals.

In their crusade to improve the quality of domestic lighting, the illumination engineers found an ally in those progressive reformers who were working to improve public health and domestic economy. In 1893, these activists created the National Household Economic Association, with a mission to spread the gospel of "domestic science," raising public awareness about the latest research on health and sanitation. They advocated more hygienic food preparation, the value of pure water and modern plumbing, the latest insights into the sources and prevention of disease, and the role of "good light in a sanitarily built house." Working through university extension programs and women's clubs, these crusaders believed that many social problems could be traced back to the average American's "gross ignorance and indifference" about the proper way to eat, wash, clean, and sleep.[33]

A key tool for creating a modern, hygienic household was strong, clear light, which not only exposed the dirt and germs lurking in dim places but created a home environment that was functional and beautiful, a living space that nurtured the body and elevated the spirit. Reformers put a particular faith in the healing and purifying powers of natural light, urging every American to pull back the curtains and take daily sunbaths. But they recognized that "the artificial-light habit grows with civilization," and so devoted much energy to promoting the proper care and use of household lamps. After shelter and clothing, as one put it, artificial light ranked as "the third most important of the necessities of civilization."[34]

By the 1920s, the authors of these household manuals began to assume that many of their readers would either have electric lights or would at least consider them a viable option. Echoing the sentiments of illumination engineers, these domestic scientists declared electric lighting the best way to create a living space that was beautiful, cheerful, and sanitary. And they advised the wise consumer to avail herself of all the comforts and conveniences that electric companies had devised—lights in hallways and closets, a mixture of indirect overhead lighting and artfully designed lamps, switches at each entrance to allow people to move in and out of electrified space, and a lighting "design" creating a distinctive atmosphere in each part of the house. In a popular 1928 home economics textbook, the professor urged students to arrange a class visit to a construction site to examine wiring plans for themselves. She advised them to study the various kinds of lamps and fixtures shown in magazine ads and to hold class discussions on "where each could be suitably used in the house." Such a lavish spread of light remained beyond the reach of working-class and rural families for some time, but a growing number of Americans were coming to see electric light as an essential part of a safe and "hygienic" household.[35]

These changes in technology produced a corresponding change in the way middle-class American families interacted once the sun went down. Some complained that since family members felt less compelled to draw together each night around a common lamp, their bonds had weakened and the art of conversation had suffered. People talked less and read more, as cheaper books and more evening light encouraged the explosive growth of what people at the time called a new "reading habit." The electrified home provided every family member with more personal space and more independence. For the first time, even younger children could be trusted alone in their rooms after dark, using switches to safely light their own way, no longer endangered by the risks of an open flame.

Pioneers in the illumination field believed that their profession would deliver even more than safer, more efficient, and more beautiful environments. Combining ancient ideas about the influence of light on the soul with the progressive reformer's impulse to shape human nature, illuminators felt themselves to be embarked on a mission of "almost sublime importance," developing artificial light's power as a tool to instill mood and create character. Poorly lit rooms produced "nervous irritation," stressing the mind and spirit and sapping an individual's moral energy. Brilliant glare had a similar enervating effect, dulling the mind and contributing to all the "mental and moral" defects of modern life. But thanks to the illuminator's art, Laurent Godinez predicted, all that was ending. American culture's "crude commercialism of yesterday" would soon yield to a growing appreciation for "the finer and more beautiful things of life" that better light could provide.[36]

Good lighting, then, was a vital tool for creating good people, a subtle form of social control that reached through the eye to instruct and shape human personality. As Godinez surveyed the new field of illumination engineering in 1911, he declared that this "movement" had grown far beyond its promoters' "expectations and hopes." The first generation of electricians had been clever but misguided "utilitarians" who cared only about delivering the most light for the least money. But now illumination engineers would craft lighting that would help people become better citizens, more optimistic, public-spirited, and sociable. It would improve the productivity and "industrial morale" of American workers while luring them after hours to enjoy all the visual pleasures of a modern consumer economy. Impressive public lighting could revitalize neighborhoods and give the poor a glimpse of beauty and hope in their difficult lives. At home, illumination promised to enrich each of the day's activities and liberate the individual by encouraging privacy and reading. In churches, modern illumination could even provide the soul with "consciousness of a sublime relationship with the Infinite."[37]

The electric light's power to accomplish all this was only enhanced by its growing invisibility to modern Americans. Artificial light became more powerful and specialized, but also more ordinary and ubiquitous, engineered into the background of daily life. Lighting designers brought theatrical effects into every public space and private room, inventing or refining languages of light that modern urban people have since come to "read" without reflecting upon them. And so the light itself, its power to create the "moods and illusions" of modern life, its central role in work and play, and its influence over the rhythm of our day becomes most "visible" on those rare occasions when the power goes off and the lights go out.[38]

By 1920, however, this technology that was doing so much to shape the texture of daily life and the sensibilities of modern Americans had not yet arrived for millions: those who lived beyond the reach of central power stations, and those city dwellers too poor or transient to pay for what remained to them the "light of the future." Here, in the place where gas and oil lamps still flickered, the centrality of electric light to a full participation in American society became most visible—painfully so to many on the dark side of this divide. Once a luxury, electric light had become a middle-class comfort, and was fast becoming an ever more important marker in a social and cultural divide between rural and urban America, the nation's past and its future.

Twelve

Rural Light

In the early 1880s, electricians had predicted that their new light would soon become "the sole illuminant of mankind." Instead, while the electric lighting business grew steadily through the early twentieth century, gaslight managed to hold its own, even in many places within reach of a central electric station. Roused by electricity's challenge, the lumbering gas industry had turned nimble, improving efficiency tenfold and introducing some inventions of its own such as electric starters and brighter incandescent burners. Electric companies scored many early victories over gas, winning contracts to light streets and parks, factories and public buildings, larger retail emporiums and the homes of the wealthy. But in the contest for the home lighting market, less than 15 percent of American households were wired for electricity by 1910, almost all of them in urban areas.[1]

As the electric industry matured, the dangers of its product had been contained but not removed, and some homeowners remained uneasy about bringing this potentially deadly force into their homes.

Though city codes and insurance rules improved electricity's safety record, a medical journal in 1900 tallied sixty "accidental electrocutions" across the country in a single day, and electrical fires caused many millions of dollars in damages every year. As one paper observed, "there is apparently no end to the different ways in which electricity can start fires." Accident victims often suffered from their own creative uses of the technology. The public was warned, for example, that it was not a good idea to try to light a cigar on the flame of an arc light or to use an incandescent lamp as a bed warmer. And many learned the hard way not to touch a live wire while sitting in a bathtub.[2]

Others held out against electricity to avoid the dust and disruption that was an inevitable part of having an older home wired—the walls and floorboards ripped open, the old gas fixtures pulled out and replaced. Cost posed the most significant barrier as gas men continued to offer the best "poor man's light." Gas stank, it rotted the furniture and soiled the walls and carpets, but it was cheap. By one estimate, a consumer lighting a home for four hours each night would spend two and a half dollars a year with an improved gas burner, but more than eighteen dollars for the same amount of incandescent electric light. And so in spite of electricity's undeniable advantages, in 1910 a leader of the gas industry could plausibly boast, "We have had the lighting market largely to ourselves." Three years later, *Scientific American* reported that "the market for electricity has only been skirted, and never really penetrated."[3]

In part, electricity's slow penetration into the home reflected the priorities of the utility companies themselves, which earned greater profits serving energy-intensive customers in the fields of manufacturing, transportation, retail, signs, and street lighting. Most wired homes still used their electric light sparingly, and only the more affluent demanded much extra current to power multiple appliances, a conservatism inspired in part by the utilities' own rate structure.

And so outside of densely populated and wealthy neighborhoods, home lighting remained a secondary concern for most electric companies.[4]

The pace of electrification accelerated in the decade after World War I. Improved incandescent bulbs and arcs replaced gas streetlamps on secondary roads, as suburban residents clamored for more light in their neighborhoods and pedestrians demanded better protection from the growing threat of automobile traffic. At the same time, architects began to design buildings that made no provision for gas lighting and builders stopped installing dual fixtures that let tenants choose between gas and electricity. Thanks to the invention of ever more efficient bulbs and more powerful distribution systems, the electric companies finally seemed poised to deliver on Edison's vow to provide an interior light that was an affordable alternative to gas.[5]

Through the 1920s, some utilities and electrical manufacturers mounted more aggressive campaigns to expand into the home lighting market, inviting customers to tour model electrified homes and showrooms of the latest electrical fixtures and sending out an army of solicitors to drum up new customers. "Electricity interests every one," an industry sales manual explained, "and is to a certain extent desired by every one." True enough, but since wiring a house was a major investment for middle-class homeowners, the salesman still needed tricks to get customers to set aside fiscal prudence and yield to the nagging tug of consumer desire. For example, if a potential customer balked at the price of a full installation, the solicitor might suggest putting in a front-porch light, which could be paid for at a flat monthly rate. Here was the thin edge of the wedge—the homeowner would soon fall in love with his front-porch lamp and the glowing message it sent to the passing world about the progressive-minded and affluent folks who lived within. Before long, the customer would agree to have the whole house wired. The door-to-door

In the 1920s, General Electric touted the luxury of lamps for every household purpose, convenient plugs, and wall switches so that the homeowner could "unroll a path of light" when entering every room.

electricity salesmen were also advised, when interviewing "the lady of the house," to let her know "what her neighbors are doing, and so play upon her social pride, insinuating, in a delicate way, that if they can afford it, she can."[6]

Getting wires into the many new homes built during the postwar building boom was much easier since these buyers saw the value of installing wires from the start. From that point on, even modest homes within reach of a power station included wiring as a matter of course. Thanks in part to easy credit and a growing market for household appliances, the number of electrified households continued to grow, with 70 percent of homes wired by 1930.

As electric light grew more pervasive in the culture, it also found its way into the American lingo. Almost from the start, speakers found the electric light a handy metaphor. Reformers vowed to shine a *piercing electric beam* on every social problem, while preachers urged the members of their flock to shed its exposing light on their sins. Temperance leader Frances Perkins seemed particularly fond of this metaphor, invoking the "electric light of Christ's love" in her war on drink.

Creative adaptation of electrical terms pushed further as light became more ubiquitous. Some unknown cartoonist first depicted a character with a shining bulb overhead and thus invented modern society's universal symbol of inventive thinking. In the 1890s an energetic and engaging person became a "live wire," and by the 1920s Americans described the opposite, a dull or boring person, as a "dim bulb." Rapidly disappearing objects had always "gone out like a light," but in the electrical age sudden changes happened at "the flip of a switch." Electric light's superiority over its older rivals became such common knowledge that it even found its way into the U.S. Army's notorious intelligence test, which required Word War I draftees to explain why electric light was better than gas.

But for the 30 percent of American families who lived beyond the reach of any central power station, essays on the glory of electric light remained purely an academic exercise, and they experienced the new technology only on their trips to town. From the start, companies offered isolated stand-alone lighting plants that promised country homeowners all the comfort and convenience enjoyed by "city gentlemen." But such systems remained "the extreme outpost of luxury," far too expensive for most rural families. A 1910 government survey of rural life estimated that electricity had arrived at only 2 percent of American farms.[7]

And so early-twentieth-century housekeeping manuals still offered detailed instruction on the use of a range of other fuels. The scientific and technological revolution of the nineteenth century had in fact produced more choices than ever for domestic lighting, as the oldest technologies remained viable, and even improved. Some people burned a gasoline mixture, inexpensive but prone to disaster. Others tried acetylene, a gas produced from calcium carbide that created a strong white light but also enough explosions to scare off many. The vast majority in rural areas relied on kerosene. In 1908, Americans bought $133 million worth, about the same price their more technologically blessed neighbors paid for electricity. While kerosene was superior to other fuels, the lamps demanded some skill to operate efficiently and a good deal of time each day cleaning mantles and trimming wicks. "The care of lamps requires so much attention and discretion," Catherine Beecher observed, "that many ladies choose to do the work themselves rather than trust it with domestics." Though more stable than other fuels, kerosene also required regular testing, since adulteration made it dangerously volatile. Popular housekeeping guides often provided several pages of tips for kerosene users who wanted to avoid burning the house down, including this dire advice: "When a lamp overflows or for any reason gets on fire, seize it and throw it out the window."[8]

Even the modern "Welsbach" gas lamp, with an incandescent mantle that many considered a viable rival to Edison's bulbs, required careful adjustment to avoid belching a smoke that could discolor the ceiling and clog the burner. A skilled housewife or domestic servant spent considerable time cleaning its chimney and needed some expertise to change its mantle, which was "exceedingly delicate, and will fall to pieces at a touch."[9]

Electricity, by contrast, demanded nothing from its customers but the timely payment of a monthly bill. Many companies still did not even require users to change their own light bulbs, but sent employees to handle the job when they broke or, more likely, grew dim and inefficient. Other utilities required customers to buy and replace their own bulbs, assuring them that the job posed no danger and required zero technical skill, so simple that "even a servant" could learn to do it. The grid hid all of its technological complexity from consumers, who tapped into an enormously powerful and expanding energy network, perhaps the greatest engineering achievement of the modern age, simply by mastering the on/off switch.[10]

Some Americans had begun to miss their pre-electrical past even before electricity arrived. As early as the 1870s, a nostalgic reaction against modern life and its "mechanical way of living" produced the first signs of what came to be known as the colonial revival. Advocates of the "House Beautiful" movement urged Americans to restore old fireplaces, minimize the use of gas in favor of the more pleasing candle and kerosene lamp, and in other ways "restore again some of the cheerful comforts of our forefathers." In the late nineteenth century, visitors to national fairs and expositions marveled at the new machines that would shape the nation's future, but they also flocked to see replicas of old New England kitchens and the vanished southern plantation house, models of various places where George Washington once slept, and antique-filled period rooms, each complemented by the flicker of candles and a cheering blaze in the hearth.

To some extent, touring these relics of the "olde tyme" served to underscore the nation's remarkable progress, making clear how much the country's technology had evolved over the course of the century. At a time of rapid and disorienting change, Americans who sought out these exhibits could take pride in their culture's past while also congratulating themselves for living in an age that left behind all the inconveniences of the tallow wick and the smoky lamp. But for many, the colonial revival offered a refuge from modern technologies that had delivered so much yet somehow failed in their promise to create a healthier and more comfortable life.[11]

Clarence Cook, a leader in the colonial revival movement, captured this yearning when he advised his readers to refrain from lighting their modern lamps for at least an hour as the sun went down, and to try instead "real candlesticks with real candles . . . casting a soft, wavering beam . . . and bringing a few moments of poetry to close the weary working-day." Just after the Civil War, the home economist Catherine Beecher had observed that "candles are used only on rare occasions," but forty years later America's candle industry was thriving. Now casting a nostalgic glow, candlesticks returned to the well-appointed home as part of a wider fad for antiques that swept the country in the early twentieth century. Guiding yourself to bed with a candle, as one put it, "makes you feel that life is a little less complex."[12]

While etiquette books praised the way candlelight flattered a woman's complexion, the economist Thorstein Veblen saw less flattering motives at work in the candle's revival. As its price dropped by the turn of the century, electric light no longer served as an emblem of high social status for the wealthy but became a more commonplace middle-class comfort. Elegant homes and expensive restaurants burned candles, Veblen suggested, because the feeble tapers had once again become a "wasteful" expenditure that only the rich could afford. "Candle-light is now softer, less distressing to well-bred eyes, than oil, gas, or electric light," he noted in 1899. "The same could not have been said thirty years ago, when candles were, or recently had been, the cheapest available light for domestic use."[13]

But the ambivalence about electricity's strong light, and the longing to experience the "pleasant associations" of a flickering flame, suggest something deeper in the culture than a hunger for conspicuous consumption. As electric light grew more pervasive, shaping the lives of city dwellers more with each passing year, some associated artificial light with an artificial life. As one critic put it, office workers in places such as New York had lost any connection to the sun and sky. They commuted to their jobs in tunnels, scurried along sidewalk canyons in the shade of skyscrapers, toiled under office lights all day, and shopped in vast emporiums where the artificial lights never went off. In the early days of electric light, social reformers had been shocked by the story of "the family under the bridge," the sweatshop workers who slept by day under the shade of the Brooklyn Bridge and worked by night under its bright arc lights. Such a topsy-turvy life, out of step with nature's rhythm, now seemed all too common, to some extent the lot of most urban workers. "Man is becoming a Mole," as one critic put it. "Nature did not equip him or intend him for such a role."[14]

America's leading psychologist, G. Stanley Hall, shared this concern about the impact of artificial light on human development and feared it might even be warping the soul. Studying the impact of light and dark on the mental and emotional development of children, Hall speculated that a new generation of Americans, growing up under artificial light, less often experienced dusk's twilight. With their rich colors and long shadows, these were the very hours that nurtured something poetic and spiritual in the human psyche, a contemplative mode of experience that looked beyond the mundane material world exposed by the strong light of day. His 1903 investigation showed that when artificial lights came on, disrupting the poetic reverie of dusk, this provoked in young people a "fevered state of mind that is marked by abandon," an unhealthy nervous excitement best cured by a trip to the country. A generation of Americans had welcomed the incandescent bulb as a beacon showing the way toward ever more civilization, but Hall and many of his fellow intellectuals feared that their culture had become too civilized for its own good, and they gave artificial light some of the blame.[15]

Edison shared in this ambivalence about the modern technological culture he had done so much to create. Starting in 1916, he took yearly camping trips with his friends Henry Ford, the tire manufacturer Harvey Firestone, and the naturalist John Burroughs. In the early twentieth century, many city-dwelling Americans felt the same urge to escape the trappings of modern life, seeking temporary shelter in a tent far from the "harsh warning of trolley gongs," the rumble of elevated trains, and "the electric lights that amaze the sight." A popular camping guide of the era asked its readers, "Is it good for men and women and children to swarm together in cities and stay there, till their instincts are so far perverted that they lose all taste for their natural element, the wide world out-of-doors?" His answer was no, of course, and the remedy tried by many "overcivilized men" was a camping vacation.[16]

And so Edison joined the growing movement of Americans who longed for something missing in their hectic urban lives, something that might be restored if they could only get "back to nature and rough it in the wilderness." "I don't want to be near electricity," Edison explained to reporters. "An old suit, an old hat, a few French novels and the fishing rod, that's all I bother with." Touring the Adirondacks, his "gypsy" band of celebrity industrialists searched for the small dirt roads where they hoped to avoid all those other touring motorists who shared this desire for a less hectic, rural way of life.[17]

Though Edison told reporters he aimed to forget the modern world, he brought much of it along, hauling enough equipment to require several cars, two trucks, and the help of four servants, including two chefs and a "laboratory expert" in charge of water and sanitation. As Edison and Ford's camping excursions became an annual summer event, the entourage grew each year, including more industrialists sampling the simple life, flocks of reporters, and even a visit from President Warren Harding in 1921. By then, the group

Thomas Edison, roughing it with his fellow "vagabonds," the naturalist John Burroughs and the industrialists Henry Ford and Harvey Firestone, West Virginia, 1918.

used specially designed camping cars and brought along a player piano and a set of Edison's electric lights for every tent. Still, Edison described his annual escape into the woods as his "feeble protest against civilization." At the close of his long career, this man who had done so much to create the technological trappings of modern life concluded that civilization itself was "only a veneer." Every man, as he put it, "deep down in his heart revolts at civilization. Every man will revert to barbarism if given half a chance."[18]

For more than forty years, commentators had been hailing incandescent light as a beacon of progress that was guiding humanity on civilization's upward path. But while rusticating and ruminating far from his laboratory, the great inventor Edison expressed some doubts that more artificial light, or any other technology, would soon drive out the darkness in human nature. Taking many of these camping trips during World War I, when industrial slaughter raged in France, and London and Paris extinguished their lights each night to avoid zeppelin attacks, Edison had good reason to doubt that modern technology was clearing an easy path to human perfection. As the United States found itself drawn into the conflict, he had quickly volunteered his inventive talents to the government's armament program, and later urged Congress to create its first permanent research laboratory dedicated to the development of weapons of war.[19]

In spite of the war's cautionary tale and his own longing to escape the artificialities and distractions of modern culture, Edison still retained much of his technological optimism. Summing up his philosophy of life in 1920, he reckoned that environment shapes human character but also that technology gives humans the power to shape that environment. To illustrate the point, he suggested that people who were deprived of artificial light would gradually sink into barbarism. But if you "put an undeveloped human being into an environment where there is artificial light," he explained, "he will improve."[20]

On the one hand, Edison decided that human evolution must be an extremely gradual process. In spite of all that science and technology had accomplished in his lifetime to improve the human condition, he believed that humanity had only just begun its quest for knowledge and moral improvement, a journey toward perfection that he estimated would take the species another fifteen thousand years to complete. On the other hand, he claimed to see immediate improvements in human character once it had been exposed to a bit of incandescent light. No systematic philosopher, Edison in his contradictory statements reflects an ambivalence over the social and ethical implications of technology shared by many ever since, those who have lived in a world shaped by so many of Edison's great inventions. Machines have simultaneously seemed like both a distraction and a concentration of human powers, the highest achievement of human nature and a fount of alienation, a tool to liberate us from nature's confines and a prison of our own making. Still inventing into his eighties, Edison concluded that the best humans could do was to keep moving. "There is hope for civilization," as he put it, "as long as man probes into the unknown and keeps experimenting."[21]

This quandary over the downside of modern machines is a concern best enjoyed by those who already have them, not those still waiting for their chance. As electrical service became far more common in American cities and towns after World War I, farm advocates pressed for the extension of electric lines to serve the millions of rural families still without power. By the mid-1930s, only one in nine American farms had electrical service. The United States produced and consumed more electricity than the rest of the world combined but, thanks in part to its vast spaces, lagged far behind other industrialized nations in providing universal access to its service—a growing disparity that quite literally left rural Americans feeling disempowered. In their magazines and newspapers, farmers read General Electric's advertisements touting electricity as a "cultural service" that offered "a finer leisure, a greater wealth, a better health, and a broader civic life." And yet here was a product that most rural families could not buy at any reasonable cost.[22]

Through the 1920s, small-scale experiments in rural electrification showcased the benefits of electric light and power on the farm, but private utilities had little incentive to make the massive investment needed to reach these potential customers. The Great Depression broke this stalemate when the Roosevelt administration made rural electrification a centerpiece of its relief programs, first through the Tennessee Valley Authority (TVA) and then with the 1936 Rural Electrification Act. These programs put the federal government into the power-generating business, revived interest in the municipal ownership of utilities, and loaned millions of dollars to farmers' cooperatives to run power lines into rural areas. Since the costly expense of rural wires could be recovered only if the new customers consumed lots of energy, the government also offered low-interest loans on plumbing and appliances and sent an army of agriculture and domestic-science experts into the field to demonstrate all the ways farm life could be improved by electricity, encouraging rural Americans to buy their electrical future on the installment plan.[23]

Just as many city and town dwellers had done a generation earlier, rural neighbors gathered to welcome electricity's arrival, marking the event with baseball games and band concerts, and in some cases a ceremonial funeral for a kerosene lamp, symbol of all the dimness and drudgery now left behind. Gone forever were evenings spent trying to "darn socks by a kerosene lamp" or milking cows in the dim shadows of "an oil lantern hanging on a nearby nail." In the first decades of electric light, urban residents had thanked and praised the entrepreneurs and electricians who had performed what seemed like a technological miracle, vaulting their towns into the modern age with the flip of a switch. A half century later, farmers felt a similar sense that they had stepped across a threshold into "modern living." The guests of honor at these ceremonies were not electricians and entrepreneurs but government

Rural Electrification Administration poster.

officials with an expertise in regional economic planning and "long-range co-operative financing." Making electric light universal was no longer a technical challenge but an economic and political one.[24]

Electric light was a welcome addition in farm homes, reduced the risk of barn fires, and in the henhouse stimulated greater egg production. Electric power proved even more transformative, as electric pumps enabled farmers to install indoor plumbing and irrigate fields, and other electric motors made many farm chores less backbreaking. Rural electrification aimed to make farming more profitable, while also helping urban workers by opening a vast new market for electrical appliances.

But the New Deal's electrical reformers aimed for something even bigger than an economic stimulus to relieve farmers and put workers back in business. They hoped to use this program to heal a cultural rift between urban and rural America that had been widening for decades, as city populations boomed, rural villages dwindled, and many farmers felt increasingly alienated from the economic and social mainstream. By the 1920s, the center of cultural authority had shifted in America; public opinion was now dominated by cosmopolitans who lampooned country "hayseeds" and dismissed parochial thinking as "small town stuff." Once proud to think of themselves as belonging to a nation of farmers, many Americans now faced an identity crisis.[25]

The New Deal's social engineers believed that rural electrification would do much to ease the burden of farm work and give rural people more access to leisure time and the conveniences of the city's modern consumer economy. Once they enjoyed access to the same light and power enjoyed by their "town cousins," country people would feel less isolated from the mainstream of American life and less tempted to leave the farm for the lure of city lights. Reformers hoped not only to stop the drain of country people into the cities but to reverse this flow, solving the chronic urban problems of crowding, crime, and pollution by offering city dwellers the chance to escape to a healthier environment without forsaking their electric lights and running water. At the same time, New Deal policy encouraged electrical modernization for all homes, urban and rural, putting in place the economic and legal framework that made bright light and high energy consumption a defining characteristic of daily life in modern America.[26]

The line between the lit and unlit world remained hard to erase—the project of rural electrification took more than the ten years first envisioned by the Rural Electrification Act, and reduced but never eliminated the technological divide between urban and rural America. But it contributed to the development of an important democratic principle: the idea that some technologies are so fundamental to the growth of human dignity and happiness that access to them could become something like an American birthright, guaranteed to all as a matter of public policy. "Electricity is no longer a luxury," as Franklin Roosevelt declared; "it is a definite necessity."[27]

Epilogue

Electric Light's Golden Jubilee

In 1929, Henry Ford organized a grand international celebration to honor Edison on the fiftieth anniversary of his first successful incandescent light bulb, bringing the inventor and honored guests to Michigan for a ceremony broadcast by radio around the world. By then, Edison's pioneering role in the field of electricity had long since been eclipsed by other scientists and engineers, many of them working in research and development labs on both sides of the Atlantic. Thanks to their accomplishments, the incandescent bulb had evolved from a headline-grabbing "miracle" to a mundane household object, even as modern bulbs provided many times more light than Edison's first lamps. In the midst of these radical changes in both his invention and the invention process, Edison's fame as the nation's greatest living inventor remained more sure than ever—he was the grandest hero in a vanishing heroic age of invention, a burst of individual achievement and technological genius that in only fifty years had laid the foundation of modern America, a transformation so rapid that Edison's accomplishments had already begun to look quaintly historic. Reflecting that change in perspective, Ford reverently gathered all that remained of Edison's Menlo Park laboratory, making an exhibit of the inventor's career a central focus of his new museum of American industry and invention in Dearborn. He even transported the topsoil from the Menlo Park grounds, though Edison did not share Ford's sentimental attachment to what he called "the same damn Jersey clay."

At the ceremony, speakers reviewed the many inventions that earned Edison the title of "founder and father of the present industrial era," but each one singled out the incandescent light as his greatest work. A half century after Edison and his rivals launched the electric lighting industry, the technology's economic and social value had grown too big to fully grasp, one speaker noted, an accomplishment that made him "one of the greatest men who ever lived." In 1879, a handful of inventors in Europe and America had toiled to create a viable light. Since then, ten billion incandescent lamps had been produced, close to a million Americans worked in some field of electrical work, and electric light and power had transformed every aspect of the way Americans worked and played. So pervasive had the light become that its impact could best be appreciated in its absence, and so the "Night of Light" ceremony included a sixty-second plunge into darkness.

During this grand celebration of Edison's career, no one bothered to dredge up the technical details about the inventor's specific contributions to the wider international effort to create a viable electric lighting system. Instead, the country celebrated the simpler truism that Edison was *the* inventor of the light bulb. As one paper put it, he was "The Man Who Lighted the World."

Having moved on from electric lighting, Edison had been devoting his last years to finding a source of rubber from native American plants—a minor success made irrelevant by the German invention of synthetic rubber. In 1892, when he lost control of his lighting companies during the General Electric merger, Edison had vowed to use his talents to create a new invention so amazing that history would forget his role in inventing the light bulb. Since then, he had enjoyed some successes, expanding his motion picture business and marketing an improved battery to power electric cars. But he had also spent many years and millions of dollars on ideas that never came to fruition, and invented nothing that rivaled the transformative power of his incandescent light.[1]

Still, Edison remained the nation's favorite authority on all things electrical, and the industry's boosters were pleased to publish his claim that after fifty years of remarkable growth, the country remained only "half lighted," with enormous potential for growth in the years ahead. Reminiscing about his pioneering days, Edison recalled that so many had once been afraid of his invention that he had been forced to "practically give it away." But thanks to steady improvements in the technology, fewer Americans each year could resist the lure of electric light—a tool of economic growth, a time-saving convenience, and, thanks to the illuminator's art, an object of beauty, something not only to *see by* but to *look at.* "Let a man see an example of better lighting," Edison explained, "and in most cases he will want it. It's simply a manifestation of human nature."[2]

Reporters loved to ask Edison for predictions about how new inventions would change life for future generations, and he loved to oblige, ignoring his own protest that "prognostication is not one of my pastimes." Contemplating the final conquest of electric light in the culture, he promised Americans a future in which the setting sun presented "no obstacles" to human activity. The grand Gilded Age promise would at last be fulfilled, as ever more powerful lighting systems would erase all distinction between night and day. Edison felt sure that this "mastery over the forces of nature" was about to liberate humanity to pursue its "greatest development."

. . .

During the broadcast of *Light's Golden Jubilee,* an orchestra played the inventor's favorite tunes from 1879 while he received congratulations sent from princes, prime ministers, business leaders, and scientific dignitaries around the world. Gerard Swope, president of General Electric, joined Edison at the speaker's table. Founded in large part on Edison's valuable lighting patents, GE now controlled 96 percent of America's incandescent lighting business, producing an annual profit of over $30 million. Other celebrities mingling in the audience included Will Rogers, John D. Rockefeller, Orville Wright, and Marie Curie, while Albert Einstein's greeting was broadcast via telephone from Berlin. Admiral Richard E. Byrd cabled from the South Pole, noting that during blizzards his team used electric light to guide men to safety, making Edison "a benefactor even to those at the very ends of the earth." Though suffering stage fright, Edison was deeply moved by a tribute that one journalist hailed as "the greatest outpouring of public sentiment and appreciation ever received by any man in the entire history of the world."[3]

The night's dramatic finale came when Edison played the lead role in a radio play designed to recreate the mythical moment when he tested his first successful carbon-filament lamp. Leaving the banquet hall, Edison walked to the reconstructed Menlo Park laboratory, his path illuminated by a few historic oil lamps and the raucous glow of Ford's nearby River Rouge automobile factory. Since the room was small, only a few select guests watched the scene—President Herbert Hoover, Ford and his son Edsel, and Francis Jehl, the lone survivor among Edison's partners in the Menlo Park experiments. After firing the dynamo, Edison connected two wires to a replica of his original lamp while the radio announcer did his best to infuse the moment with suspense. "Will it light?" he asked in a breathless whisper. "Will it burn?" Of course it did. As the an-

As Henry Ford and Francis Jehl look on, Edison reinvents the electric light for an international radio audience, 1929.

nouncer informed the radio audience of this "triumphant climax," cities across the country lit specially designed lamps, and bells tolled in Philadelphia's Independence Hall. Meanwhile, Edison fled the room, flustered by the public spotlight and pleading for a nap.

Later that evening, President Hoover gave the celebration's keynote speech. As a former engineer, he was more qualified than most other politicians to evaluate Edison's legacy. The ceremony came just days before that month's terrifying world stock market crash. As the world economy hovered on the brink of disaster, Hoover declared that Edison's life story "gives renewed confidence that our institutions hold open the door of opportunity to all who would enter." Nothing affirmed the economic value and democratic promise of industrial capitalism like the genius of its great inventors, and no person embodied Hoover's faith in American individualism better than Thomas Edison.

While honoring all that Edison had accomplished through his personal genius and hard work, Hoover singled out for praise his role in creating "the modern method of invention." What Edison had shown the world at Menlo Park a half century earlier was not just a new light bulb but the power of well-funded, organized technological research. "In earlier times," the president explained, "mechanical invention had been the infrequent and haphazard product of genius in the woodshed." Edison demonstrated that the future belonged to those who toiled in the "great laboratories of both pure and applied science," the system of corporate research he had helped to create and that had long since eclipsed his role on the frontier of electrical discovery. Though most of these university-trained scientists and engineers remained anonymous, by 1929 their work was producing more inventions than ever, a vital engine of economic and intellectual progress that expanded and enriched human life in countless ways.

In a similar fashion, Hoover suggested that the impact of Edison's lights far surpassed anything that one man could have envisioned in 1879. The great inventor had aimed to liberate people from their gas and oil lamps by providing a better and cheaper light. But far beyond that, the humble bulb that left Edison's laboratory and made its way in the world had been transformed by the creativity of others to serve "an infinite variety of unexpected uses":

> It enables us to postpone our spectacles for a few years longer; it has made reading in bed infinitely more comfortable; by merely pushing a button we have introduced the element of surprise in dealing with burglars; the goblins that lived in dark corners and under the bed have now been driven to the outdoors; evil deeds which inhabit the dark have been driven back into the farthest retreats of the night; it enables the doctor to peer into the recesses of our insides; it substitutes for the hot water bottle in aches and pains; it enables our towns and cities to clothe themselves in gaiety by night, no matter how sad their appearance may be by day. And by all its multitude uses it has lengthened the hours of our active lives, decreased our fears, replaced the dark with good cheer, increased our safety, decreased our toil, and enabled us to read the type in the telephone book. It has become the friend of man and child.[4]

In this, Hoover recognized what most other speakers had missed that day: that what Edison and his rival inventors had done fifty years earlier was to release upon the world a technology with enormous potential, one with far too many possibilities for any one person to anticipate or create. Edison conceded as much. "When I laid the foundation of the electrical industry," he told a reporter during the festivities, "I did not dream that it would grow to its present proportions. Its development has been a source of amazement to me."[5]

Hoover's playful list of the light's various applications captured some of these "amazing" and unexpected consequences of Edison's invention, though his tally was far from complete. To it we should add the light's role in expanding human knowledge of the deep sea and the microscopic world, of caves and polar darkness. The light inspired creative adaptations by architects, urban planners, lighting designers, sign makers, window dressers, and theater artists. And we should remember the untold number of inventions and social conventions that others developed in order to fully realize the light's potential: the engineering standards, insurance guidelines, consumer protections, and utility regulations. Hoover noted that light made Americans more productive, less afraid and vulnerable, but might have also mentioned its role in creating other distinctive markers of modern life—our frenetic pace and long hours, our assumption that any barrier imposed by nature can be overturned or ignored in the name of economic efficiency, the exhilarating and disorienting effects of electric mass marketing and retail, and the experience of living, working, and playing in spaces carefully engineered by illuminators to induce a proper mood.

Though Hoover captured just a part of the electric light's enormous economic, social, and cultural impact, he deserves credit for acknowledging that Edison, for all his greatness, was only the foremost among many, some known and others long forgotten, who had played some part in inventing the light bulb and thus in creating this essential part of modern culture. Edison recognized this more than anyone, and offered the fullest tribute that evening to all those who had joined him in building a world of electric light. In a voice quavering with emotion, the world's greatest inventor told his global audience, "I would be embarrassed at the honors that are being heaped on me were it not for the fact that in honoring me you are also honoring that vast army of thinkers and workers of the past, and those who are carrying on; without whom my own work would have gone for nothing. If I have spurred men to greater effort and if our work has widened the horizon of man's understanding even a little and given a measure of happiness to the world, I am content."

After giving his brief remarks, Edison collapsed and had to be taken to Ford's house for several days of bed rest. "I am tired of all the glory," he complained. "I want to get back to work."

Acknowledgments

While research is often a solitary experience, this topic has led me to the company of many, and the chance to talk with them about electric light over the past few years has made this book a pleasure to write. My interest in the topic began with lively conversations with students in my graduate seminars at the University of Tennessee, and I enjoyed much good advice and support from my colleagues in the history department: thanks especially to Steve Ash, Denise Phillips, Tom Burman, Lynn Sacco, Tom Coens, Tom Chaffin, and Dan Feller. Keith Lyon provided excellent help with the research, tracked down sources with great energy and a discerning eye, and was fine company as we followed electric light down so many surprising alleys. Anne Bridges and James McKee also provided helpful research assistance.

I am very grateful for the early encouragement and support I received from Christine Heyrmann, John Demos, Joyce Seltzer, Alan Rutenberg, and Patrick Allitt. Steve Ash, Paul Israel, Steve

Whitaker, and Patrick Allitt provided keen readings of the manuscript, as did Scott Moyers and Mally Anderson at the Penguin Press. Each shared insights that made this book better.

The American Council of Learned Societies provided a generous fellowship that afforded me the time to complete this manuscript. My time in the archives of the New-York Historical Society was supported by a grant from the Gilder Lehrman Institute of American History. A Huntington Library fellowship provided access to its outstanding Dibner collection on the history of electricity, and I am grateful to Daniel Walker Howe and many of the visiting fellows for engaging conversations in one of academia's nicest groves. Most insightful and most fun, Holly Clayson generously shared her own interesting research on electric light's early days. Two research grants from the Winterthur Museum and Library allowed me to spend time in that rich archive, and provided the chance for lively conversations about American material culture. I am particularly grateful to fellowship coordinator Rosemary Krill, who did so much to make my visits to Delaware productive and enjoyable, to librarians Emily Guthrie and Helena Richardson, estate historian Maggie Lidz, and to my fellow fellows—especially Laurie Churchman, Louisa Iarocci, and Jennifer Carlquist. The University of Tennessee also provided research awards that allowed me to visit archives in the United States and England.

My friend and mentor Don Coonley provided much moral support as I began this project—as a photographer and filmmaker, he understood more than most the lure of light. I regret that because of his untimely death I will not have the pleasure of hearing his thoughts on this book.

Thanks, as ever, to my wife, Lauren, for her love, companionship, and keen questions, and her patience during my long absences and short weekends.

ACKNOWLEDGMENTS

For more than half a century now, my parents have been a boundless source of love, friendship, and encouragement. I dedicate this book to them—my mother, Jane, the writer, and my father, Ernest II, the engineer who tried so patiently to show me how the world works.

Notes

INTRODUCTION: INVENTING EDISON

1 *Journal of Experimental Social Psychology* 46 (2010): 696–700.
2 Irwin Unger, *These United States: The Questions of Our Past*, 2nd ed. (Upper Saddle River, NJ: Prentice Hall, 2003), 415.
3 Paul Israel, *Edison: A Life of Invention* (New York: Wiley, 1998), chapter 10.
4 Ibid., 167; Jill Jonnes, *Empires of Light: Edison, Tesla, Westinghouse, and the Race to Electrify the World* (New York: Random House, 2003), 66–67.
5 On the lighting revolution prior to the development of electric light, particularly the impact of gas lighting in urban areas, see Wolfgang Schivelbusch, *Disenchanted Night: The Industrialization of Light in the Nineteenth Century* (Berkeley and Los Angeles: University of California Press, 1995), and on the American context, Peter C. Baldwin, *In the Watches of the Night: Life in the Nocturnal City, 1820–1930* (Chicago: University of Chicago Press, 2012).

ONE: INVENTING ELECTRIC LIGHT

1 *Dickens' Dictionary of Paris* (1882), 84.
2 *New York Tribune*, September 30, 1881.
3 Most of the modern criticisms of these technologies are as old as the technologies themselves. For a review of some criticisms of Western technology in this period, see Michael Adas, *Machines as the Measure of Men: Science, Technology, and Ideologies of Western Dominance* (Ithaca, NY: Cornell University Press, 1989), chapter 6.
4 *Scotsman*, August 12, 1881.
5 Friedrich Ratzel, *Sketches of Urban and Cultural Life in North America*, trans. and ed. Stewart A. Stehlin (New Brunswick, NJ: Rutgers University Press, 1988), 3–4; Charles Reade cited in Erastus O. Haven, *The National Hand-Book of American Progress* (1876).
6 Ralph Waldo Emerson, *Works and Days* (1857).
7 On Emerson's evolving views about the meaning of American invention, see John Kasson, *Civilizing the Machine: Technology and Republican Values in America, 1776–1900* (New York: Grossman, 1976).
8 Harold Platt, *Electric City: Energy and the Growth of the Chicago Area, 1880–1930* (Chicago: University of Chicago Press, 1991), 28.

9 Peter C. Baldwin, *In the Watches of the Night: Life in the Nocturnal City, 1820–1930* (Chicago: University of Chicago Press, 2012), 16.

10 Patricia Fara, *An Entertainment for Angels: Electricity in the Enlightenment* (New York: Columbia University Press, 2002); Michael Brian Schiffer, *Draw the Lightning Down: Benjamin Franklin and Electrical Technology in the Age of Enlightenment* (Berkeley and Los Angeles: University of California Press, 2003); James Delbourgo, *A Most Amazing Scene of Wonders: Electricity and Enlightenment in Early America* (Cambridge, MA: Harvard University Press, 2006); Jill Jonnes, *Empires of Light: Edison, Tesla, Westinghouse, and the Race to Electrify the World* (New York: Random House, 2003), 18–29.

11 Richard Holmes, *Age of Wonder: The Romantic Generation and the Discovery of the Beauty and Terror of Science* (New York: Vintage, 2010), 295–96.

12 Brian Bowers, *Lengthening the Day: A History of Lighting Technology* (New York: Oxford University Press, 1998), 63–65; Schiffer, *Draw the Lightning Down*, 228–32.

13 *Proceedings of the American Academy of Arts and Sciences*, May 1894, 415; *The Anglo American, a Journal of Literature, News, Politics, the Drama*, December 23, 1843, 2, 9.

14 Wolfgang Schivelbusch, *Disenchanted Night: The Industrialization of Light in the Nineteenth Century* (Berkeley and Los Angeles: University of California Press, 1995), 52–57; *Milwaukee Daily Sentinel*, September 21, 1876; "The Electric Light: How It Is Now Being Utilized in Paris and Throughout France," *Chicago Tribune*, February 7, 1879; on Faraday, see Jonnes, *Empires of Light*, 38–44.

15 *New York Tribune*, June 27, 1878.

16 *Chicago Tribune*, March 19, 1877, March 27, 1878; *Journal of the Franklin Institute*, July 1, 1849; *Chicago Tribune*, January 30, 1879, April 1, 1878.

17 *Scientific American*, May 24, 1879.

18 *Cleveland News*, April 14, 1928.

19 *Cleveland Herald*, April 28, 1879; Marie Gilchrist, Charles Francis Brush, unpublished manuscript, 1935, Case Western Reserve University Special Collections; *Chicago Tribune*, April 30, 1879.

20 Earl Hamer, "A Reminiscence about the First Lighting in Wabash, Indiana" (1929), transcript at Wabash County Historical Museum; Linda Simon, *Dark Light: Electricity and Anxiety from the Telegraph to the X-Ray* (Orlando: Harcourt, 2005), 80–81.

21 *Christian Union*, July 23, 1879; *Washington Post*, August 4, 1879.

22 On Palmer House: Platt, *The Electric City*, 22; on circus: *Atlanta Constitution*, October 2, 1879.

23 *Chicago Tribune*, April 28, 1879.

24 Robert Louis Stevenson, "A Plea for Gas Lamps," *Virginibus Puerisque and Other Papers* (London: C. Kegan Paul & Company, 1881), 288.

25 *Boston Globe*, July 30, 1882.

26 *Indianapolis Sentinel*, September 17, 1882.

27 Robert Hammond, *The Electric Light in Our Home* (1884); *Electrical World*, August 3, 1883; *Detroit Free Press*, November 21, 1880.

28 Robert Friedel and Paul Israel, *Edison's Electric Light: The Art of Invention* (Baltimore: Johns Hopkins University Press, 2010), 5–8.

29 *District School Journal of the State of New York* (1840–1852), August 1846, 5–7; "Report on the International Exhibition," 160; *Los Angeles Times*,

February 1, 1893; Roscoe Scott, "Evolution of the Lamp," *Transactions* (1914); on Moses Farmer, see Dirk Struik, *Yankee Science in the Making: Science and Engineering in the Making from Colonial Times to the Civil War* (Mineola, NY: Dover, 1992), 332; Goebel in Schivelbusch, *Disenchanted Night*, 58.

30 Paul Israel, *Edison: A Life of Invention* (New York: Wiley, 1998), 164–69; Friedel and Israel, *Edison's Electric Light*, 4–9.

31 *New York Sun*, September 16, 1878; Jonnes, *Empires of Light*, 55–56.

32 Friedel and Israel, *Edison's Electric Light*, 8–9.

33 *New York Tribune*, March 3, 1879.

34 Friedel and Israel, *Edison's Electric Light*, 10, 24.

35 *Atlanta Constitution*, January 18, 1881. Maxim, who had grown up in the backwoods of Maine and went on to a remarkable career inventing everything from mousetraps to machine guns, had installed a lighting system of his own design in the New York Post Office, the Park Avenue Hotel, and in some Wall Street offices. He grew tired of hearing crowds gather around, asking him if this was "Edison's famous light." "As Edison had never made a lamp up to that time," he recalled years later, "I was annoyed." Hiram Maxim, *My Life* (1915), 130; on Edison's self-regulating platinum filament bulb, see Friedel and Israel, *Edison's Electric Light*, chapter 1.

36 Jonnes, *Empires of Light*, 57.

37 *New Haven Register*, December 30, 1879; *Daily Inter Ocean*, May 17, 1882; *Chicago Tribune*, May 6, 1879.

38 *Chicago Tribune*, April 15, May 22, and May 16, 1879; *Chicago Tribune*, March 5, 1879.

39 *New York Times*, March 1, 1881.

40 Matthew Josephson, *Edison: A Biography* (New York: McGraw-Hill, 1959), 224–25; Francis Jehl, *Menlo Park Reminiscences* (Dearborn Park, MI: Edison Institute, 1937), 410–30. On Edison's decision to return to the use of carbon, and the Menlo Park demonstration, see Friedel and Israel, *Edison's Electric Light*, chapter 4, and Jonnes, *Empires of Light*, 58–67.

41 Israel, *Edison: A Life of Invention*, 187–88.

42 *Washington Post*, January 1, 1880.

43 *New York World*, December 22, 1879; *Electrical World*, January 15, 1881; *New York Times*, January 16, 1880; Frank Leonard Pope, *Evolution of the Electric Incandescent Lamp* (1889), 30; Jehl, *Menlo Park Reminiscences*, 230–32.

44 Israel, *Edison: A Life of Invention*, 188–89.

45 Andrew Hickenlooper, *A Memoir* (typescript), Cincinnati Historical Society; interview with Hickenlooper, October 27, 1878, clipping in Cincinnati Historical Society.

46 *New York Sun*, June 14, 1880; *Electrical World*, March 19, 1881.

47 "Electric Lighting: A Lecture," Joseph Swan, October 20, 1880, Newcastle; *Newcastle Daily News*, June 10, 1881; Charles Bazerman, *The Languages of Edison's Light* (Cambridge, MA: MIT Press, 2002), 187–88; Graeme Gooday, *Domesticating Electricity* (London: Pickering & Chatto, 2008), 163–64.

48 *Electrical World*, April 30, 1881.

49 On Sawyer's challenge to Edison's invention, see Jehl, *Menlo Park Reminiscences*, 396–97.

50 Bazerman, *The Languages of Edison's Light*, 209–10; Josephson, *Edison: A Biography*, 228; Malcolm MacLaren, *The Rise of the Electrical Industry During the Nineteenth Century* (Princeton: Princeton University Press, 1943), 74;

Charles Wrege and Ronald Greenwood, "William E. Sawyer and the Rise and Fall of America's First Incandescent Light Company, 1878–1881," *Business and Economic History*, 2nd ser., vol. 13 (1984): 31–48.

51 Edison sent more than machines to Paris to make his case, as two of his trusted partners spent weeks schmoozing with Europe's electrical elite, and in more than one case bought out the services of influential French journalists. Robert Fox, "Thomas Edison's Parisian Campaign: Incandescent Lighting and the Hidden Face of Technology Transfer," *Annals of Science* 53 (1997): 157–93. Friedel and Israel, *Edison's Electric Light,* 180.

52 Bazerman, *The Languages of Edison's Light*, 202.

53 Preece in *Van Nostrand's Engineering Magazine*, February 1, 1882; his comments on Edison's character cited in E. C. Baker, *Sir William Preece, F.R.S.: Victorian Engineer Extraordinary* (London: Hutchinson, 1976), 157; Bazerman, *The Languages of Edison's Light*, 214.

54 Thomas Hughes, *Networks of Power: Electrification in Western Society, 1880–1930* (Baltimore: Johns Hopkins University Press, 1993), 20–21; Israel, *Edison: A Life of Invention*, chapter 10.

55 *Times* (London), March 15, 1881; *Chicago Tribune*, August 16, 1881.

56 *Standard*, October 3, 1881; Henry Edmunds, "Reminiscences of a Pioneer," December 25, 1919, reprinted from *M and C Apprentices Magazine* (Glasgow), clippings in Joseph Swan papers, Newcastle upon Tyne; Gooday, *Domesticating Electricity*, 94–95.

57 "Gas-Light in Paris," *Daily Evening Bulletin*, December 22, 1881; *Newcastle Daily Chronicle*, April 14, 1881.

58 William Edward Sawyer, *Electric Lighting by Incandescence* (1882), 184.

TWO: CIVIC LIGHT

1 Harold Platt, *The Electric City: Energy and Growth of the Chicago Area, 1880–1930* (Chicago: University of Chicago Press, 1991), 22.

2 *New York Times*, December 21, 1880.

3 *San Jose Mercury*, December 25, 1881; *Los Angeles Herald*, December 16, 1881; *Los Angeles Times*, February 2 and August 5, 1882; Eddy Feldman, *The Art of Street Lighting in Los Angeles* (Los Angeles: Dawson's Bookshop, 1972), 23.

4 *Los Angeles Times*, April 28 and August 10, 1883; Feldman, *Art of Street Lighting in Los Angeles*, chapter 3.

5 *Atlanta Constitution*, November 29 and December 14, 1883, July 30, May 25, and June 1, 1884.

6 Paul de Rousiers, *American Life* (1891), 131.

7 *Electrical World*, December 27, 1884.

8 Ibid., February 20, 1886.

9 Ibid., April 19, 1883, April 20, 1886; on the problem of sexual violence in American cities, see Peter C. Baldwin, *In the Watches of the Night: Life in the Nocturnal City, 1820–1930* (Chicago: University of Chicago Press, 2012), 176–78.

10 *Milwaukee Journal,* September 26, 1894, citing *New York Sun;* on gaslight as a magnet for prostitution, see Andreas Bluhm and Louise Lippincott, *Light!: The Industrial Age, 1750–1900* (Amsterdam: Van Gogh Museum, 2000), 212; Baldwin, *In the Watches of the Night*, chapter 2, 160. As Baldwin shows, some evidence suggests that well-lit streets actually attracted pickpockets and prostitutes, drawn to the crowds gathered there.

11 Baldwin, *In the Watches of the Night*, 27–33; *Electrical Review*, May 30, 1885.
12 Platt, *The Electric City*, 23; *Electrical Age*, August 1904.
13 *Electrical World*, May 15, 1886; Alan Trachtenberg, *The Incorporation of America: Culture and Society in the Gilded Age* (New York: Hill & Wang, 1982), 105.
14 *Electrical World*, November 17, 1887; "Arc Lights to Repel Lovers," *New York Times*, August 20, 1905; *Electrical World*, January 3, 1884; *New Haven Register*, January 10 and April 10, 1885.
15 *Electrical World*, March 15, 1885; Platt, *The Electric City*, 29.
16 *Electrical World*, December 21, 1889; *New York Times*, December 28, 1888.
17 *Electrical World*, June 26, 1886, on Laramie; *Electrical World*, October 13, 1883, quote from *Fargo Argus*.
18 Billy Nye, *Remarks by Bill Nye* (1887).
19 *Electrical World*, October 18, 1884.
20 Chris Otter, *The Victorian Eye: A Political History of Light and Vision in Britain, 1800–1910* (Chicago: University of Chicago Press, 2008), 226–28.
21 *Detroit Free Press*, October 10, 1880.
22 *Chicago Times* cited in *Los Angeles Times*, January 25, 1882.
23 *Detroit Free Press*, July 23 and November 16, 1884.
24 Ibid., June 24, 1886.
25 Ibid., August 1, 1884, July 3, 1885; *Electrical World*, June 6, 1885.
26 *Electrical World*, March 31, 1888.
27 *San Francisco Call*, September 22, 1907.
28 Platt, *The Electric City*, 38–39.
29 *St. Louis Post-Dispatch*, August 17, 1890; *New York Tribune*, May 7, 1890; *Baltimore Sun*, June 23, 1890; *New York Times*, July 7, 1887; *Chicago Tribune*, February 28, 1882; Tiffany in *Electrical World*, April 12, 1884; John Lewis, *A Treatise on the Law of Eminent Domain in the United States* (1888), 810.
30 Platt, *The Electric City*, 46–47.

THREE: CREATIVE DESTRUCTION: EDISON AND THE GAS COMPANIES

1 Matthew Josephson, *Edison: A Biography* (New York: McGraw-Hill, 1959), 262–63; Robert Friedel and Paul Israel, *Edison's Electric Light: The Art of Invention* (Baltimore: Johns Hopkins University Press, 2010), 183–87.
2 Paul Israel, *Edison: A Life of Invention* (New York: Wiley, 1998), chapter 11, provides a valuable summary of Edison's first central station. See also Jill Jonnes, *Empires of Light: Edison, Tesla, Westinghouse, and the Race to Electrify the World* (New York: Random House, 2003), 76–85.
3 *New York Herald*, *New York Times*, *New York Tribune*, September 5, 1882.
4 On the skilled maintenance required by gas lighting, see Sarah Milan, "Refracting the Gaselier: Understanding the Victorian Responses to Domestic Gas Lighting," in *Domestic Space: Reading the Nineteenth-Century Interior*, ed. Inga Bryden and Janet Floyd (Manchester: Manchester University Press, 1999).
5 Randall E. Stross, *The Wizard of Menlo Park: How Thomas Alva Edison Invented the Modern World* (New York: Crown, 2008), 129–31; Edwin Burrows and Mike Wallace, *Gotham: A History of New York City to 1898* (New York: Oxford University Press, 1998); Jonnes, *Empires of Light*, 3–15.
6 *Electrical World*, August 15, 1886.
7 Harold Platt, *The Electric City: Energy and the Growth of the Chicago Area,*

1880–1930 (Chicago: University of Chicago Press, 1991), 25–28; *New Orleans Picayune*, September 26, 1882; *The Papers of Thomas A. Edison*, vol. 6, *Electrifying New York and Abroad, April 1881–March 1883*, ed. Paul B. Israel, Louis Carlat, and David Hochfelder (Baltimore: Johns Hopkins University Press, 2007), xxiii; while Edison pioneered the central station model, for a number of years most of his customers bought the smaller, stand-alone systems.

8 *Detroit Free Press*, November 13, 1878, provides a thorough overview of criticism of gas companies; see also *Detroit Free Press*, December 25, 1879; Platt, *The Electric City*, 12–14. The rocky reception for gaslight in England through the nineteenth century is discussed in Milan, "Refracting the Gaselier."

9 *Electrical World*, November 15, 1884, July 4, 1886.

10 For example, *Chicago Tribune*, October 16, 1881; *New York Times*, January 24 and October 22, 1890.

11 *Boston Globe*, October 31, 1895; *Medical News*, May 19, 1888; *Chicago Tribune*, September 1, 1889.

12 *New York Times*, January 3, 1890; *New York Tribune*, January 8, 1880.

13 *American Medical Journal* (1888), 182; *Youth's Companion*, January 26, 1882; *Albany Law Journal*, December 29, 1888.

14 *Scientific American*, January 13, 1883; Platt, *The Electric City*, 42–43; *New York Times*, December 4, 12, 1889.

15 *Electrical World*, July 26 and November 8, 1884, June 13, 1885.

16 Ibid., October 13, 1883.

17 Ibid., March 30 and August 24, 1889; *Progressive Age*, October 1, 1889.

18 "Views of Dr. Edwards," *Baltimore Sun*, December 13, 1889. The American tendency to build things cheaply, using hasty designs and inferior products, has been used to explain the country's rapid embrace of new technologies—the obsolescence of any design is assumed from the start, and this imposes fewer barriers for the creative destruction needed to replace older systems with new and improved ones. See H. J. Habakkuk, *American and British Technology in the Nineteenth Century: The Search for Labour-Saving Inventions* (Cambridge: Cambridge University Press, 1962), 87–92.

19 British electric lighting discussed in E. C. Baker, *Sir William Preece, F.R.S.: Victorian Engineer Extraordinary* (London: Hutchinson, 1976), 9, 253; Preece cited in *Electrical World*, December 27, 1884.

20 *Electrical World*, June 23, 1888; Preece in *Electrical World*, December 27, 1884.

FOUR: WORK LIGHT

1 Helen Campbell, *Darkness and Daylight* (1892), 272–73; Helen Campbell, *New York Tribune*, January 2, 1887.

2 *Daily National Intelligencer*, November 2, 1865.

3 On night work as part of a larger effort to increase the pace of production in New England textile mills, see Philip Foner and David Roediger, *Our Own Time: A History of American Labor and the Working Day* (New York: Greenwood Press, 1989), 50–51.

4 "An Unnoticed Increase in Mill Capacity," *The Timberman*, October 29, 1887; Ford cited in Foner and Roediger, *Our Own Time*, 191; David Nye, *Electrifying America: Social Meanings of a New Technology, 1880–1940* (Cambridge, MA: MIT Press, 1992), chapter 1; *Cotton, Wool, and Iron*, May 12, 1883; Peter C.

Baldwin, *In the Watches of the Night: Life in the Nocturnal City, 1820–1930* (Chicago: University of Chicago Press, 2012), 125–30.

5 *Spectator* cited in *Chicago Tribune*, October 28, 1878.

6 *Electrical World*, July 26, 1884.

7 Otto Mayr and Robert C. Post, eds., *Yankee Enterprise: The Rise of the American System of Manufactures* (Washington, DC: Smithsonian Institution Press, 1981), 178–79.

8 Nye, *Electrifying America*, 191–92; Patricia Hills, *Turn-of-the-Century America: Paintings, Graphics, Photographs, 1890–1910* (New York: Whitney Museum of American Art, 1977), 89–90; Philip Meggs and Alston W. Purvis, *Meggs' History of Graphic Design* (Hoboken, NJ: Wiley, 2005), chapter 9; Robert Jay, *The Trade Card in Nineteenth-Century America* (Columbia: University of Missouri Press, 1987), 36.

9 *Electrical World*, July 21, 1883, January 12, 1884; Baldwin, *In the Watches of the Night*, 114–16.

10 *Chicago Tribune*, January 3, 1879.

11 *Scientific American*, December 21, 1878, August 28, 1880.

12 See also Wolfgang Schivelbusch, *The Railway Journey: The Industrialization of Time and Space in the 19th Century* (Berkeley and Los Angeles: University of California Press, 1987).

13 Thomas J. Schlereth, *Victorian America: Transformations in Everyday Life, 1876–1915* (New York: Harper Perennial, 1992), 22; *Electrical World*, July 12, 1890.

14 Another strategy for night trains in the antebellum period was the use of "pilot engines" that went ahead of the main train, searching for obstructions on the dark track. In a history of southern rail travel, one historian concludes that most travelers preferred instead to disrupt their journey and stop at a hotel for the night. Eugene Alvarez, *Travel on Southern Antebellum Railroads, 1828–1860* (Tuscaloosa: University of Alabama Press, 1974), 81.

15 John H. White, *A History of the American Locomotive: Its Development, 1830–1880* (Mineola, NY: Dover, 1980); *Los Angeles Times*, February 6, 1887; *Railway Age Gazette*, June 20, 1910, 1650; *Electrical World*, September 27, 1890.

16 Benjamin H. Barrows, *The Evolution of Artificial Light: From a Pine Knot to the Pintsch Light* (1893); *Scientific American*, April 9, 1887; *New York Times*, September 30, 1887; T. Clarke, *The American Railway* (1889), 226; *Electrical World*, September 7, 1889.

17 Julian Ralph, *Dixie, or Southern Scenes and Sketches* (1896), 104–5; *Electrical World*, March 15, 1890.

18 Mark Twain, *Life on the Mississippi* (1883), 225, 448, 513.

19 Joseph H. Appel, *The Business Biography of John Wanamaker, Founder and Builder* (New York: Macmillan, 1930), 102–3; Herbert Adams Gibbons, *John Wanamaker* (New York: Harper & Brothers, 1926), vol. 1, 216–17.

20 *Boston Globe*, November 22, 1880.

21 Samuel Terry, *How to Keep a Store, Embodying the Conclusions of Thirty Years of Experience in Merchandising* (1887), 113–14.

22 Baldwin, *In the Watches of the Night*, 104–14.

23 Campbell, *Darkness and Daylight*, 257–58; Baldwin, *In the Watches of the Night*, 49–53.

24 *Chicago Tribune*, January 11, 1885.

25 Roediger and Foner, *Our Own Time*, 58; Wolfgang Schivelbusch, *Disenchanted Night: The Industrialization of Light in the Nineteenth Century* (Berkeley and Los Angeles: University of California Press, 1995), 8–9; Baldwin, *In the Watches of the Night*, 130–37.
26 Foner and Roediger, *Our Own Time*, 58; Baldwin, *In the Watches of the Night*, 186–90, 194–200.
27 *Los Angeles Times*, September 24, 1880; "The Flesh and Blood of Children Coined into Dollars," *New Century Path*, May 31, 1903; Baldwin, *In the Watches of the Night*, 135–36.
28 *Georgia Weekly Telegraph*, December 10, 1882.
29 *New Hampshire Statesman*, September 18, 1868; *Alienist and Neurologist*, April 1, 1887; *St. Louis Globe-Democrat*, April 22, 1881.
30 Alan Trachtenberg, *The Incorporation of America: Culture and Society in the Gilded Age* (New York: Hill & Wang, 1982), citing Charles Francis Adams Jr. on the railroads, 45.

FIVE: LEISURE LIGHT

1 Robert Louis Stevenson, "Plea for Gas Lamps," *Virginibus Puerisque and Other Papers* (London: C. Kegan Paul & Company, 1881); Jonathan Bourne and Vanessa Brett, *Lighting in the Domestic Interior: Renaissance to Art Nouveau* (London: Sotheby Parke Bernet, 1991), 193.
2 *Chicago Tribune*, September 27, 1892.
3 Wolfgang Schivelbusch, *Disenchanted Night: The Industrialization of Light in the Nineteenth Century* (Berkeley and Los Angeles: University of California Press, 1995), 50–51.
4 *New York Times*, September 29, 1891; *Electrical World*, November 14, 1891.
5 *Electrical World*, March 6, 1886; Schivelbusch, "The Stage," *Disenchanted Night*; *Electrical World*, August 25, 1883; "History of Footlights," *Brooklyn Eagle*, January 12, 1896; attempts to use gas and limelight in this way are discussed in Frederick Penzel, *Theatre Lighting Before Electricity* (Middletown, CT: Wesleyan University Press, 1978), 60–63, 73–74.
6 *Baltimore Sun*, September 3, 1884.
7 David Nye, *Electrifying America: Social Meanings of a New Technology* (Cambridge, MA: MIT Press, 1992), 32–35; *Milwaukee Republican-Sentinel*, December 26, 1882.
8 *St. Louis Post-Dispatch*, July 2, 1888.
9 Julian Ralph, *Harper's Chicago and the World's Fair* (1893).
10 Nye, *Electrifying America*, 35; on Coney Island, see John Kasson, *Amusing the Million: Coney Island at the Turn of the Century* (New York: Hill & Wang, 1978), and Woody Register, *The Kid of Coney Island: Fred Thompson and the Rise of American Amusements* (New York: Oxford University Press, 2003); *Chicago Tribune*, August 27, 1893.
11 *Puck*, June 25, 1884.
12 *New York Times*, July 16, 1886.
13 Ibid., July 10, 1887; Erastus Wiman, *The Gospel of Relaxation* (1887).
14 *Electrical World*, July 2, 1887.
15 "The Electric Fountain at Lincoln Park," *Harper's Weekly*, October 3, 1891.
16 *New York Times*, August 8, 1880; *Puck*, August 3, 1880; "The Evolution of the Modern Amusement Park," *Street Railway Journal*, January 25, 1908; Gary

Kyriazi, *The Great American Amusement Parks: A Pictorial History* (New York: Citadel, 1976), 69.

17 Kasson, *Amusing the Million*, 43–46; Register, *The Kid of Coney Island*, 132.

18 Charles Belmont Davis, "The Renaissance of Coney," *Outing Magazine*, August 1906.

19 "The Development of Summer Lighting," *Electrical Age*, April 1, 1904.

20 *Electrical World*, May 19, 1888; Rollin Hartt, *The People at Play* (1909), 123.

21 *Los Angeles Times*, August 26, 1888.

22 *Police Gazette*, June 23, 1883.

23 David Pietrusza, *Lights On!: The Wild Century-Long Saga of Night Baseball* (Lanham, MD: Scarecrow Press, 1997); "Night Lighting for Outdoor Sports," *Bulletin, National Lamp Works of General Electric Co. Engineering Department*, November 5, 1925.

24 *St. Louis Post-Dispatch*, June 26, 1892; *Harper's Weekly*, August 18, 1894.

25 Luther Stieringer, "From Christmas Tree to Pan-American," *Electrical World*, August 24, 1901.

26 The idea was obvious, and others claimed to be the first to use electric lights, including Edward Johnson, an Edison Company vice president from New York who electrified his tree in 1882. See Penne Restad, *Christmas in America: A History* (New York: Oxford University Press, 1995), 114; Anthony and Peter Miall, *The Victorian Christmas Book* (New York: Pantheon, 1978), 55–59; *Electrical World*, January 12, 1884, January 3, 1885; Leigh Eric Schmidt, *Consumer Rites: The Buying and Selling of American Holidays* (Princeton: Princeton University Press, 1995), 159–69.

27 *New York Times*, January 1, 1907.

28 *New York Times* and *New York Tribune*, November 1, 1884.

29 William T. Elsing, "Life in New York Tenement Houses, as Seen by a City Missionary," *Scribner's Magazine*, June 1892; on the Metropolitan Museum, *Electrical World*, May 31, 1890.

30 Rodney Welch, "The Farmer's Changed Condition," *The Forum*, February 1891; *An Inquiry into the Causes of Agricultural Depression in New York State* (1895); "The Tendency of Men to Live in Cities," *Journal of Social Science* (1895), 8; *New York Times*, September 3, 1895. For a useful overview of evening entertainments and cultural activities available in the nineteenth-century city, see Peter C. Baldwin, *In the Watches of the Night: Life in the Nocturnal City, 1820–1930* (Chicago: University of Chicago Press, 2012), chapter 4.

31 Charles Loring Brace, *The Dangerous Classes of New York* (1872); S. L. Loomis, *Modern Cities and Their Religious Problems* (1887); Thomas Bender, *The Unfinished City: New York and the Metropolitan Idea* (New York: New York University Press, 2003), 166–80.

SIX: INVENTIVE NATION

1 Jane Mork Gibson, "The International Electrical Exhibition of 1884: A Landmark for the Electrical Engineer," *IEEE Transactions on Education*, August 1980; on the superiority of American inventiveness, see *Saturday Evening Post*, May 4, 1881; James Dredte, ed., *Electric Illumination*, vol. 2 (London, 1885). A German perspective is noted in *Science*, January 23, 1885; William J. Hammer, "The Franklin Institute Exhibition of 1884," typescript in William J. Ham-

mer Scientific Collection, Smithsonian Institution; *General Report of the Chairman of the Committee on Exhibitions* (1885).

2 Bruce Sinclair, *Philadelphia's Philosopher Mechanics: A History of the Franklin Institute, 1824–1865* (Baltimore: Johns Hopkins University Press, 1974); "Elihu Thomson," *Science*, October 17, 1924; A. Michal McMahon, *The Making of a Profession: A Century of Electrical Engineering in America* (New York: IEEE Press, 1984), 21–22; on the British model of using institutions to promote science, see David Knight, *Public Understanding of Science: A History of Communicating Scientific Ideas* (New York: Routledge, 2006), chapter 3.

3 *The Electrician and Electrical Engineer*, October 1884.

4 *American Catholic Quarterly Review*, October 1884.

5 William Preece cited in *Electrical World*, December 27, 1884.

6 Henry Schroeder, "History of Incandescent Lamp Manufacture," *General Electric Review*, September 1911.

7 *Electricity in Its Relation to the Mechanical Engineer*, November 28, 1887; John Kasson, *Civilizing the Machine: Technology and Republican Values in America, 1776–1900* (New York: Grossman, 1976), 183–66; on the wider context of technological inventiveness in European societies, see Robert Friedel, *A Culture of Improvement: Technology and the Western Millennium* (Cambridge, MA: MIT Press, 2007).

8 "The Port of New York," *Frank Leslie's Popular Monthly*, June 1883; Randall E. Stross, *The Wizard of Menlo Park: How Thomas Alva Edison Invented the Modern World* (New York: Crown, 2008), 42–44; Kasson, *Civilizing the Machine*, 41, 48–49.

9 "The Romance of Invention," *Parry's Monthly Magazine*, June 1887; J. B. McClure, ed., *Edison and His Inventions* (1898); Wyn Wachhorst, *Thomas Alva Edison: An American Myth* (Cambridge, MA: MIT Press, 1981), 52–86; Phillip Hubert, *Inventors* (1896), 248–57; "Talks with Edison," *Harper's*, February 1890; Edison cited in Jill Jonnes, *Empires of Light: Edison, Tesla, Westinghouse, and the Race to Electrify the World* (New York: Random House, 2003), 85.

10 *Electrical World*, June 7, 1884.

11 *Popular Science*, November 1877; Thomas P. Hughes, *American Genesis: A Century of Invention and Technological Enthusiasm* (Chicago: University of Chicago Press, 1989), chapter 1.

12 *Telegraphist*, October 1, 1884; *Potter's American Monthly*, July 1881; Kasson, *Civilizing the Machine*, 21–22.

13 Nathaniel Shaler, *Thoughts on the Nature of Intellectual Property and Its Importance to the State* (1877), 23; B. Zorina Khan, *The Democratization of Invention* (Cambridge: Cambridge University Press, 2005), 5; "Imitation and Invention," *San Francisco Chronicle*, July 21, 1889.

14 "Talks with Edison," *Harper's*, February 1890.

15 *Times* (London), August 22, 1878. This discussion simplifies or ignores other incentives for inventiveness that have been identified by modern economic historians. For additional information on the economic sources of inventiveness, see Kenneth Sokoloff, "Inventive Activity in Early Industrial America: Evidence from the Patent Records, 1790–1846," *Journal of Economic History* (December 1988): 1813–50. Sokoloff's data confirms the nineteenth-century view that New England, and more widely the Northeast, produced the largest amount of inventive activity, stimulated by growing urban demand and access to markets through

transportation networks. See also Brooke Hindle, *Emulation and Invention* (New York: New York University Press, 1981), which emphasizes the importance of cultural factors that encouraged American creativity.

16 Cited in *Potter's American Monthly*, July 1881.

17 On the wider American context for these educational efforts, see Friedel, *A Culture of Improvement*, chapter 20; Hughes, *American Genesis*, 70. On British versions of this public interest in science and scientific education, see Knight, *Public Understanding of Science*, 40–42; *Electrical World*, October 26, 1889; classes at the Mechanical Institute, Cincinnati, are discussed in *Electrical Review*, April 18, 1885.

18 *Scientific American*, March 9, 1878.

19 *History of the Electrical Art in the United States Patent Office, C. J. Kintner, Lecture Delivered to Franklin Institute . . . May 15, 1886*; see also *Scientific American*, June 8, 1878. Commissioner Charles E. Mitchell, *Paper Read at the Centennial Celebration of the Beginning of the Second Century of the American Patent System at Washington, April 8, 1891.*

20 Benjamin Franklin, *Autobiography and Other Writings* (Oxford: Oxford University Press, 1999), 120.

21 Levin H. Campbell, *The Patent System in the United States: A History* (1891) provides a good explanation of the origin and evolution of the patent system through the 1836 reform; Floyd Vaughan, *The United States Patent System: Legal and Economic Conflicts in American Patent History* (Westport, CT: Greenwood Press, 1972), 18–19; Stacy Jones, *The Patent Office* (New York: Praeger, 1971), 5–7, 9; Silvio Bedini, in *The Smithsonian Book of Invention* (Washington, DC: Smithsonian Institution, 1978); David Noble, *America by Design: Science, Technology, and the Rise of Corporate Capitalism* (New York: Oxford University Press/Knopf, 1979), 84–87.

22 B. Zorina Kahn, *The Democratization of Invention: Patents and Copyrights in American Economic Development, 1790–1920* (Cambridge: Cambridge University Press, 1995), 3, 55. On women patentees, *New Orleans Picayune*, July 24, 1870; Kahn, chapter 5; Anne Macdonald, *Feminine Ingenuity: Women and Invention in America* (New York: Ballantine, 1992); for estimates on the number, and a summary of the wide range of patents held by women in the nineteenth century, see Autumn Stanley, *Mothers and Daughters of Invention: Notes for a Revised History of Technology* (New Brunswick, NJ: Rutgers University Press, 1995); on patent fees, Kahn, 7, 31, 54. "Thinking into things" is from Mitchell, *Paper Read at the Centennial Celebration of the Beginning of the Second Century of the American Patent System at Washington.*

23 "Our Patent System and What We Owe It," *Scribner's*, November 1878; on the need by various inventors to "invent around" electric light patents, see Israel, *Edison: A Life of Invention*, 316–17.

24 Earl W. Hayter, "The Patent System and Agrarian Discontent, 1875–1888," *Mississippi Valley Historical Review* (June 1947), 59–82; see *Electrical World*, July 1891, on Edison patent resolution; *Electrical World*, August 15, 1891; Israel, *Edison: A Life of Invention*, 317–18.

25 *Scientific American*, March 9, 1878.

26 *New York Tribune*, May 22, 1883; Japanese envoy cited in Kahn, *The Democratization of Invention*, 13. Both Britain and Germany also reformed their patent systems on America's "more liberal basis." Mitchell, *Paper Read at the Centen-*

nial Celebration of the Beginning of the Second Century of the American Patent System at Washington; Michael R. Auslin, *Pacific Cosmopolitans: A Cultural History of U.S.-Japan Relations* (Cambridge, MA: Harvard University Press, 2011), 57.

27 Harry Laidler, *Socialism in Thought and Action* (1920), 211–13; Edward Bellamy, *Looking Backward, 2000–1887* (1887), 226.

28 William Mallock, *A Critical Examination of Socialism* (1907), 169; Max Hirsch, *Democracy versus Socialism* (1901).

29 H. J. Habakkuk, *American and British Technology in the Nineteenth Century: The Search for Labour-Saving Inventions* (Cambridge: Cambridge University Press, 1962), 190–91.

30 *Electrical World*, October 13, 1883.

31 On the relationship between great inventors and less important "innovators," see Kahn, *The Democratization of Invention*, chapter 7; Alan Trachtenberg, *The Incorporation of America: Culture and Society in the Gilded Age* (New York: Hill & Wang, 1982), 55; "The Copper Industry," *Wall Street Journal*, October 21, 1906; Henry Schroeder, "History of Incandescent Lamp Manufacture" (1911), 8; Arthur A. Bright Jr., *The Electric-Lamp Industry: Technological Change and Economic Development from 1800 to 1947* (New York: Macmillan, 1949), 212.

32 *Electrical Review*, April 11, 1885; *Lewiston (ME) Journal* cited in *Electrical World*, May 28, 1887.

33 *Electrical World*, January 8, 1887; H. G. Prout, "Some Relations of the Engineer to Society," September 1906, included in *Addresses to Engineering Students* (1912); *Electrical World*, November 15, 1890.

SEVEN: LOOKING AT INVENTIONS, INVENTING NEW WAYS OF LOOKING

1 "Death of Mr. Greeley," *Popular Science Monthly*, January 1873; *Popular Science Monthly*, May 1872, 113.

2 *Godey's Lady's Book*, January 1888; J. B. McClure, ed., *Edison and His Inventions* (1898); on the evolution of women's magazines to include more scientific and practical subjects, see "Literature for Women," *The Critic*, August 10, 1889.

3 Henry Meigs, *An Address, On the Subject of Agriculture and Horticulture, Oct 9th 1845* (1845); Brooke Hindle, *Emulation and Invention* (New York: New York University Press, 1981).

4 Alan Trachtenberg, *The Incorporation of America: Culture and Society in the Gilded Age* (New York: Hill & Wang, 1982), 41; H. C. Westervelt, *American Progress: An Address at the Eighteenth Annual Fair of the American Institute* (1845); John Kasson, *Civilizing the Machine: Technology and Republican Values in America, 1776–1900* (New York: Grossman, 1976), 139–42.

5 *A Complete Check List of Household Lights Patented in the United States, 1792–1862*, typescript copy compiled by Howard G. Hubbard, Winterthur Library; Charles Leib, "Remember the Ladies: Nineteenth Century Women Lighting Patentees," *The Rushlight*, March 2008.

6 *Electrical Review*, June 14, 1884; on the British counterpart to these exhibitions, see David Knight, *Public Understanding of Science: A History of Communicating Scientific Ideas* (New York: Routledge, 2006), chapter 7.

7 American Institute Fair scrapbook, October 3, 1883, New-York Historical Society; *Scientific American*, October 29, 1881; Kasson, *Civilizing the Machine*, 142–48.

8 Frankland Jannus, "The Protection of Electrical Inventions," *Electrical World*, May 30, 1885.
9 *Electrical World*, January 8, 1887.
10 *American Machinist*, clipping in American Institute papers, 1897, New-York Historical Society.
11 Nathaniel Shaler, *Thoughts on the Nature of Intellectual Property* (1877), 26.
12 Helen M. Rozwadowski, *Fathoming the Ocean: The Discovery and Exploration of the Deep Sea* (Cambridge, MA: Belknap Press of Harvard University, 2005), 73, 214–15; *Electrical Review*, January 17, 1885; *Scientific American*, September 22, 1883.
13 Joseph Thorndike, ed., *Mysteries of the Deep* (New York: American Heritage, 1980), 298–301; *Electrical Engineer*, July 13, 1888; *Scientific American*, September 28, 1889.
14 *New York Tribune*, May 5, 1882; *Nature*, May 17, 1894, Fridtjof Nansen, *Farthest North* (1897); Roland Huntford, *Nansen: The Explorer as Hero* (London: Duckworth, 1997), chapters 41–43; *The Graphic*, March 11, 1882.
15 *The American Medical Bi-Weekly*, May 25, 1878; William Benjamin Carpenter, *The Microscope and Its Revelations* (1875), 390; *Buffalo Surgical and Medical Journal* (1883), 495.
16 *Annals of Anatomy and Surgery* (1883), 123; *The Epitome: A Monthly Retrospect of American Practical Medicine* (1883), 203; *Electrical World*, February 20, 1886; *Baltimore Sun*, October 4, 1881.
17 *Munsey's Magazine*, November 1893; Frederick Cartwright, *The Development of Modern Surgery* (London: Barker, 1967), 285–86; Harvey Graham, *The Story of Surgery* (New York: Doubleday, 1939), 367.
18 John Harvey Kellogg, *Neurasthenia* (Battle Creek, MI: Good Health Publishing Co., 1916), 84–85; Robert C. Fuller, *Alternative Medicine and American Religious Life* (New York: Oxford University Press, 1989), 30–34; Henry Cattell cited in Ira M. Rutkow, MD, *American Surgery: An Illustrated History* (Philadelphia: Lippincott Williams & Wilkins, 1998), 227.
19 *Journal of the American Medical Association*, June 22, 1895; *Medical Review* (1894), 490; J. H. Kellogg, *The Home Book of Modern Medicine* (1914), 685–86; "Bathing in Electric Light," *Philadelphia Inquirer*, June 27, 1897.
20 *Therapeutics of the Electric Light*, undated publication, Winterthur Library.
21 J. H. Kellogg, *Light Therapeutics: A Practical Manual of Phototherapy for the Student and Practitioner* (1910).
22 *Literary Digest*, December 1, 1900. The arc light provided more serious medicine, and so was used more sparingly. Its blast of heat was used by Kellogg and others to stimulate the blood, rouse the appetite, and "impart the ability to sleep."

EIGHT: INVENTING A PROFESSION

1 *Scientific American*, August 6, 1887.
2 *Electrical World*, May 17, 1890; *New York Times*, January 26, 1890.
3 *Electrical World*, March 30, 1889.
4 A nice summary of many of these tasks is found in Vincent Stephen, *Wrinkles in Electric Lighting* (1885); *Electrical Age*, June 1891.
5 *Electrical World*, July 21, 1883, April 26, 1890.
6 National Electric Light Association cited in *Electrical World*, August 22, 1885.
7 National Electric Light Association, *Proceedings* (1890), 17.

8 Ibid., 20; *Electrical World*, February 15, 1890; *Harper's*, February 1890.
9 *Boston Globe*, January 21, 1890.
10 *Harper's Weekly*, December 14, 1889.
11 *Boston Globe*, December 10, 1889.
12 For the view of fire underwriters on the fire, see *Electrical World*, December 21, 1889, citing *Boston Herald*; *American Gas Light Journal*, August 4, 1890.
13 *New York Times*, October 12, 1889.
14 *Harper's Weekly*, December 14, 1889.
15 *St. Louis Post-Dispatch*, January 1 and 7, November 25, 1890.
16 *Harper's Weekly*, December 10, 1887, April 27, 1889; Harold Platt, *The Electric City: Energy and the Growth of the Chicago Area, 1880–1930* (Chicago: University of Chicago Press, 1991), 42–43; *Electrical World*, February 28, 1885; on Chicago, see *Scientific American*, October 11, 1884; F. H. Whipple, *Municipal Lighting* (1889), 177–78, 171.
17 *New York Times*, September 21, 1890; on street explosions, *New York Times*, January 19, 1890; *Electrical World*, November 2, 1889, citing *New York Press.*
18 See, for example, *New York Tribune*, January 16, 1890; Whipple, *Municipal Lighting*, 208.
19 *Electrical World*, October 26, 1889; *New York Times*, December 22, 1889, May 19, 1890.
20 *New York Tribune*, October 13, 1889; Florence Wischnewetzky, "A Decade of Retrogression," *The Arena*, August 1891; *Harper's Weekly*, January 4, 1890.
21 T. A. Edison, *North American Review*, and cited in *Electrical World*, November 2, 1889.
22 Paul Israel, *Edison: A Life of Invention* (New York: Wiley, 1998), 198; Matthew Josephson, *Edison: A Biography* (New York: McGraw-Hill, 1959), 345–50; *Boston Journal*, July 12, 1882; on Westinghouse and the AC system, see Jill Jonnes, *Empires of Light: Edison, Tesla, Westinghouse, and the Race to Electrify the World* (New York: Random House, 2003), 133–39; Israel, *Edison: A Life of Invention*, 329–31; "A Warning from the Edison Electric Co." (New York: Edison Electric Company, 1888).
23 *New York Tribune*, December 22, 1889; *Electrical World*, January 5, 1889; Jonnes, *Empires of Light*, chapters 7–8. In its war on AC power, Edison attempted to discredit the more powerful force by advocating its use in the electric chair, a project that came to fruition with the grisly electrocution of convicted murderer William Kemmler at New York's Auburn Penitentiary in August 1890. See Jonnes, *Empires of Light,* chapter 8, and Mark Essig, *Edison and the Electric Chair: A Story of Light and Death* (New York: Walker, 2004).
24 "The Electric Light Muddle," *New York Tribune*, December 28, 1889; *Evening Post* cited in *Electrical World*, December 28, 1889; *Harper's New Monthly Magazine*, February 1890.
25 *New York Times*, October 29, 1889; *Electrical World*, December 21, 28, 1889; *New York Sun* cited in *Electrical World*, January 4, 1890.
26 *Electrical World*, January 31, 1891.
27 H. W. Pope, "How Our Paths May Be Made Paths of Peace: A Paper Read at the National Electric Light Association," February 1890.
28 *New York Times*, June 11, 1881; Dr. Amory cited in *Electrical World*, March 1, 1890. *Electrical World*, November 1, 1890.
29 On municipal ownership, see Daniel Rodgers, *Atlantic Crossings: Social Politics in a Progressive Age* (Cambridge, MA: Belknap Press of Harvard University, 1998), chapter 4.

30 *Municipal Monopolies* (1899), 174.

31 Victor Rosewater, "Municipal Control of Electric Lighting," *The Independent*, November 10, 1892; "The Problem of Municipal Government," *The Sun*, August 15, 1883; Alan Trachtenberg, *The Incorporation of America: Culture and Society in the Gilded Age* (New York: Hill & Wang, 1982), 107.

32 Edward Bellamy, *Talks on Nationalism* (Chicago: Peerage Press, 1938), 125; John Kasson, *Civilizing the Machine: Technology and Republican Values in America, 1776–1900* (New York: Grossman, 1976), 191–202.

33 Edward Bellamy, "Progress of Nationalism in the United States," *North American Review*, June 1892, 742–52.

34 "Other Forms of Public Control," *American Economic Association Publications*, July–September 1891.

35 Whipple, *Municipal Lighting*, 33.

36 "Disparity in Prices of Electric Lighting," *Electrical World*, July 4, 1891.

37 "Municipal Lighting," *Advance Club Leaflets No. 2* (1891), 93; an in-depth look at the politics of utility reform in Denver and Kansas City can be found in Mark Rose, *Cities of Light and Heat: Domesticating Gas and Electricity in Urban America* (University Park: Pennsylvania State University Press, 1995), chapters 1–2.

38 M. J. Francisco, reporting at National Electric Light Association, 1890, and *Electrical World*, August 30, 1890.

39 "Theoretical Basis of Municipal Ownership," *American Economic Association Publications*, July–September 1891; *American Gas Light Journal*, March 31, 1890.

40 Edward Bemis, *Municipal Monopolies* (1899), 638.

41 Laurence Gronlund, *The New Economy: A Peaceable Solution to the Social Problem* (1907), chapter 10; *Electric Age*, February 15, 1890. On British policy on overhead wires, see Chris Otter, *The Victorian Eye: A Political History of Light and Vision in Britain, 1800–1910* (Chicago: University of Chicago Press, 2008), 242–43.

42 *Electrical World*, October 19, 1889.

43 "The Future of Electricity and Gas," *The Eclectic Magazine of Foreign Literature*, January 1885; *Medical News*, May 19, 1888; Thomas Lockwood, "Electrical Notes of a Trans-Atlantic Trip," *Electrical World*, October 19, 1889. And the French were worse, in spite of their early lead in the field and their capital's long reputation as the "City of Light." French law demanded "a license for any and every piece of wiring," the Americans noted with scorn, "and an inspection of all electrical installations." Even putting in a doorbell, one scoffed, would draw the scrutiny of a panel of French bureaucrats. In 1889, the French continued to press their claim to lead the world in electrification, hosting another international exhibition of technology that featured the Eiffel Tower, lit each evening with spotlights. Visiting American electricians could not help but admire the sight of this controversial but powerful symbol of that city's passion for electrified modernity. But when they wandered a few blocks farther into the city, they were surprised to find how dark most of it remained. There is "no electric street lighting in Paris," one American electrician reported home, less that he would expect to see "at the street corner of even the smallest towns of the States."

German electrician Arthur Wilke likewise complained in 1893 that "the Germans especially require that electricity appear only if perfectly dressed for the occasion. Electricity has a much easier life in America. There it is liked, even if the wire is bare and the light shines from a primitive wooden post. . . .

Electricity may not do that in Germany; here it must show itself to the public only in its Sunday dress, otherwise the political and aesthetic police will arrive." Wilke cited in Andreas Bluhm and Louise Lippincott, *Light!: The Industrial Age, 1750–1900* (Amsterdam: Van Gogh Museum, 2000), 30.

44 *Scientific American*, February 4, 1882.

45 Charles J. H. Woodbury, *The Fire Protection of Mills* (1882); *Scientific American*, February 4, 1882. While insurance companies took the lead in formulating electrical codes in the United States, Britain's earliest codes were developed by a professional and learned society for electrical engineering, the Society of Telegraph Engineers and of Electricians. See "Rules and Regulations for the Prevention of Fire Risks Arising from Electric Lighting," *Proceedings of the Institution of Electrical Engineers*, May 1882; *Electrical World*, October 17, 1885.

46 Mark Tebeau, *Eating Smoke: Fire in Urban America, 1800–1950* (Baltimore: Johns Hopkins University Press, 2003), 251–55; Harry Chase Brearley and Daniel N. Handy, *The History of the National Board of Fire Underwriters* (1916), 81–82.

47 "The Superstitious Fear of Electric Wires," *The Chronicle*, 122; *Electrical World*, February 1 and May 31, 1890, January 17, 1891; "Wires and Their Danger," *New York Tribune*, January 20, 1889.

48 *Scientific American*, June 17, 1893; Harry Chase Brearley, *A Symbol of Safety* (1923), chapters 4, 16; *American Architect*, August 22, 1903; Sara Wermiel, *The Fireproof Building: Technology and Public Safety in the Nineteenth-Century American City* (Baltimore: Johns Hopkins University Press, 2000), 135; H. Roger Grant, *Insurance Reform: Consumer Action in the Progressive Era* (Ames: Iowa State University Press, 1979), 131–32; for a wider discussion of fire insurance reform in this period, see chapter 4 of that book; Brearley cited in Daniel B. Klein, ed., *Reputation: Studies in the Voluntary Elicitation of Good Conduct* (Ann Arbor: University of Michigan Press, 1997), 75–84.

49 Walker cited in *Boston Herald* and *Electrical World*, December 21, 1889; *Electrical World*, January 19, 1889; Israel, *Edison: A Life of Invention*, 223–24; Thomas Hughes, *Networks of Power: Electrification in Western Society, 1880–1930* (Baltimore: Johns Hopkins University Press, 1993), 143.

50 David Noble, *America by Design: Science, Technology, and the Rise of Corporate Capitalism* (New York: Knopf, 1977), chapter 6; Alfred D. Chandler, *The Visible Hand: The Managerial Revolution in American Business* (Cambridge, MA: Belknap Press of Harvard University, 1977), 426–33; Hughes, *Networks of Power*, 163–65; Nye, *Electrifying America*, 170–73; Israel, *Edison: A Life of Invention*, 336–37; A. Michal McMahon, *The Making of a Profession: A Century of Electrical Engineering in America* (New York: IEEE Press, 1984), 33–59.

51 Israel, *Edison: A Life of Invention*, 260; Matthew Josephson, *Edison, A Biography* (New York: McGraw-Hill, 1959), 331–37; Hughes, *Networks of Power*, chapter 4.

52 Hughes, *Networks of Power*, 140–41, 160–63, 166; Thomas Hughes, *American Genesis: A Century of Invention and Technological Enthusiasm* (Chicago: University of Chicago Press, 1989), 159–75; Josephson, *Edison: A Biography*, 333, 360–61; Jonnes, *Empires of Light*, provides a full account of the rivalry between the advocates of direct and alternating current.

53 Patrick Joseph McGrath, *Scientists, Business, and the State, 1890–1960* (Chapel Hill: University of North Carolina Press, 2001), chapter 1; Hughes, *Networks of Power*, 145–60.

54 Jonnes, *Empires of Light*, 347–49.

55 McMahon, *The Making of a Profession*; Grace Palladino, *Dreams of Dignity, Workers of Vision: A History of the International Brotherhood of Electrical Workers* (Washington, DC: International Brotherhood of Electrical Workers, 1991), 5–6.

56 "Linemen's Negligence," *Saint Louis Post-Dispatch*, January 3, 1891; Palladino, *Dreams of Dignity*, 5–6.

57 Elmer Warner, "Practical Suggestions to Electric Light Wiremen," *Electrical World*, January 31, 1891.

58 "Popularizing Electrical Information," *Electrical World*, July 27, 1889; *Detroit Free Press* cited in *Electrical World*, March 20, 1886; *Electrical World*, June 21, 1890.

59 Albert Scheible, "The Electricity of the Public Schools," *Electrical World*, July 19 and 26, 1890, March 22, 1890.

60 *Electrical World*, March 5, 1887, July 19 and August 9, 1890.

NINE: THE LIGHT OF CIVILIZATION

1 Mark Twain, *Life on the Mississippi* (1883), 452.

2 *Norfolk Virginian*, cited in *Electrical World*, April 14, 1885; *St. Louis Globe-Democrat*, May 12, 1882; *Frank Leslie's Illustrated Newspaper*, May 2, 1885; *San Francisco Daily Bulletin*, May 10, 1886.

3 *Chicago Tribune*, March 6, 1898.

4 H. L. Mencken, *Happy Days* (New York: Knopf, 1940), 66. The bug is called Belostomatidae, or *Belostoma*, and is discussed in Leland Howard, *The Insect Book* (1901); *St. Louis Globe-Democrat* cited in *New York Times*, June 29, 1885; *Baltimore Sun*, September 19, 1898; *Electrical World*, July 26, 1890.

5 *Forest and Stream: A Journal of Outdoor Life*, November 1, 1883.

6 *American Catholic Quarterly Review*, April 1892.

7 *Chicago Tribune*, October 9, 1903; "The 'Animal' Furniture Fad," *Current Literature*, November 1896.

8 *New York Tribune*, August 22, 1883; *Baltimore Sun*, March 4, 1884.

9 A good summary of this argument is found in Matthew Luckiesh, *Artificial Light: Its Influence upon Civilization* (1920), chapters 1–3.

10 See, for example, Otis Tufton Mason, *The Origins of Invention: A Study of Industry Among Primitive Peoples* (1895), 106–8; Robert Hammond, *The Electric Light in Our Homes* (1884); E. L. Lomax, *The Evolution of Artificial Light* (1893); Roscoe Scott, "Evolution of the Lamp," *Transactions* (1914); Walter Hough, "The Lamp of the Eskimo," *From the Report of the U.S. National Museum* (1896), 1028.

11 Cited in *Electrical World*, July 11, 1886; Rayvon Fouche, *Black Inventors in the Age of Segregation: Granville T. Woods, Lewis H. Latimer & Shelby J. Davidson* (Baltimore: Johns Hopkins University Press, 2003), chapter 1. See, for example, "After the Lynchers," *Chicago Daily*, June 4, 1893; "An Ex-Slave Owner Speaks," *The Public*, November 28, 1903.

12 Bolivia in *Literary Digest*, September 9, 1893; Tehran in *Electrical World*, September 11, 1886. Some of the more objective advocates of "battlefield electricity" conceded that even a disciplined army of European soldiers would likely falter in the face of a strobe attack. See *Electricity in Warfare, Presented to Franklin Institute*, November 13, 1885.

13 "Signs of Promise in Mexico," *The Church at Home and Abroad*, March 1892; *Electrical World*, April 2, 1887; *Public Opinion*, May 28, 1887.

14 Matthew Frye Jacobson, *Barbarian Virtues: The United States Encounters Foreign Peoples at Home and Abroad, 1876–1917* (New York: Hill & Wang, 2001), 138–41; Michael Adas, *Machines as the Measure of Men: Science, Technology, and Ideologies of Western Dominance* (Ithaca: Cornell University Press, 1989), especially chapter 5.

15 On views of the Chinese, by contrast, see Jacobson, *Barbarian Virtues*, 31–38; "Science Prophesies the Future of the Race," *Scientific American*, March 3, 1877; *Electrical World*, May 14, 1887; Frank Carpenter, "The New Japan," *New York Times*, November 5, 1888.

16 *The Deseret Weekly*, May 25, 1895; Daniel E. Bender, *American Abyss: Savagery and Civilization in the Age of Industry* (Ithaca: Cornell University Press, 2009), 92–95; Adas, *Machines as the Measure of Men*, 357–65.

17 On this genre, see the introduction in Sam Moskowitz, ed., *Science Fiction by Gaslight: A History and Anthology of Science Fiction in the Popular Magazines, 1891–1911* (Cleveland: World Publishing Co., 1968); Paul Fayter, "Strange Worlds of Space and Time: Late Victorian Science and Science Fiction," in *Victorian Science in Context*, ed. Bernard Lightman (Oxford: Clarendon, 1997).

18 Brooks Landon, *Science Fiction After 1900: From Steam Man to the Stars* (Woodbridge, CT: Twayne, 1997), 40–50, on the Frank Reade series and the genre of boy inventor stories known as "Edisonade."

19 "Noname" [Luis Senarens], *The Electric Man: Or, Frank Reade, Jr. in Australia*, serialized in *Boys of New York*, October 10, 1886.

20 Sasha Archibald, "Harnessing Niagara Falls," *Chance*, Fall 2005.

21 Eric Davis, "Representations of the Middle East at American World Fairs, 1876–1904," in *The United States and the Middle East: Cultural Encounters*, YCIAS Working Paper Series, vol. 5 (New Haven: Yale Center for International and Area Studies, 2002); official catalog and guide book to the Pan-American Exposition (1901).

22 Robert W. Rydell, *All the World's a Fair: Visions of Empire at American International Exhibitions, 1876–1916* (Chicago: University of Chicago Press, 1987), chapter 5; David Nye, *Electrifying America: Social Meanings of a New Technology, 1880–1940* (Cambridge, MA: MIT Press, 1992), 35–36; Kerry S. Grant, *The Rainbow City: Celebrating Light, Color and Architecture at the Pan-American Exposition, Buffalo 1901* (Buffalo, NY: Canisius College Press, 2001).

23 Paul Israel, *Edison: A Life of Invention* (New York: Wiley, 2000), 410–21.

24 Grant, *The Rainbow City*, 64.

25 Leo Tolstoy, *What Is to Be Done?* (1899), 367.

26 Victor Yarros, "The Decline of Tolstoi's Philosophy," *The Chautauquan*, March 1894.

27 Charles Morris, *Civilization: An Historical Review of Its Elements*, vol. 2 (1890), 22; *Boston Journal of Chemistry*, cited in John W. Hanson, *Wonders of the Nineteenth Century* (1900), 21.

28 *Electrical World*, January 8, 1887; Lieutenant Colonel Elsdale, "Scientific Problems of the Future," *Littell's Living Age*, April 28, 1894; Edward P. Thompson, *How to Make Inventions: Or, Inventing as a Science and an Art* (1893), 181; Hanson, *Wonders of the Nineteenth Century*, 425; G. H. Babcock, "Electricity in Its Relation to the Mechanical Engineer," address at the American Institute of Mechanical Engineering (1887).

29 *Electrical World*, July 3, 1886; Israel, *Edison: A Life of Invention*, 310; G. Babcock, "Electricity in Its Relation to the Mechanical Engineer."

30 *Electrical World*, November 15, 1890; Titus Keiper Smith, *Altruria* (1895),

106; Louis Bell, *The Art of Illumination* (1902), 137; Matthew Luckiesh, *Artificial Light*, 147–48. For utopian visions of a future phosphorescent light, see W. S. Harris, *Life in a Thousand Worlds* (1905); H. G. Prout, "Some Relations of the Engineer to Society," September 1906; *Electrical World*, November 15, 1890; "Great American Industries," *Harper's New Monthly Magazine*, 1896.

31 Smith, *Altruria*, 31; see also Henry Olerich, *A Cityless and Countryless World* (1893); John Kasson, *Civilizing the Machine: Technology and Republican Values in America, 1776–1900* (New York: Grossman, 1976), chapter 5.

32 Paul Devinne, *The Day of Prosperity: A Vision of the Century to Come* (1902), 55–56.

TEN: EXUBERANCE AND ORDER

1 Charles Mulford Robinson, *Modern Civic Art: Or, the City Made Beautiful* (1904), 145.

2 *Electrical World*, December 20, 1884; *New York Times*, July 7, 1910; "Electric Sign Monstrosities," *Scientific American*, September 24, 1910, cited in Edward Rossell, "Compelling Vision: From Electric Light to Illumination Engineering, 1880–1940" (PhD dissertation, University of California, Berkeley, 1998), 148; Ross cited in William Leach, *Land of Desire: Merchants, Power, and the Rise of a New American Culture* (New York: Vintage, 1994), 49; Simeon Strunsky, *Belshazzar Court: Or Village Life in New York City* (1914), 48.

3 *The Nation*, September 16, 1909, cited in Steven Conn and Max Page, ed., *Building the Nation* (Philadelphia: University of Pennsylvania Press, 2003); Rebecca Zurier, Robert W. Snyder, and Virginia McCord Mecklenburg, *Metropolitan Lives: The Ashcan Artists and Their New York, 1897–1917* (New York: Norton, 1995); Marianne Doezema, *George Bellows and Urban America* (Washington, DC: National Museum of American Art/New York: Norton, 1995); David Shi, *Facing Facts: Realism in American Thought and Culture, 1850–1920* (New York: Oxford University Press, 1996), chapter 12; Rebecca Zurier, *Picturing the City: Urban Vision and the Ashcan School* (Berkeley and Los Angeles: University of California Press, 2006), 49.

4 G. Glen Gould, "Where There Is No Vision," *Art World*, January 1917.

5 "The Destruction of Niagara," *Littell's Living Age*, August 11, 1883; Ginger Strand, *Inventing Niagara: Beauty, Power and Lies* (New York: Simon & Schuster, 2009), 142–44.

6 Jonathan Baxter Harrison, *The Condition of Niagara Falls; and the Measures Needed to Preserve Them* (1882), 47–50.

7 William Irwin, *The New Niagara: Tourism, Technology, and the Landscape of Niagara Falls* (University Park: Pennsylvania State University Press, 1996), chapter 3; "Illumination of Niagara Falls," *General Electric Review*, February 1908, 115–19; Frederick A. Talbot, *Electrical Wonders of the World*, vol. 1 (1921), 67–69, discusses renewed attempts to light the falls.

8 On the role of William J. Hammer, an Edison employee, in developing these early signs, see "Brief Outline of Electric Sign History and Development," *Signs of the Times*, June 1916, and William J. Hammer, "Electricity and Some Things That Can Be Done with It: illustrated lecture by Hammer, YMCA Christian Union Bldg, 19 Feb 1887," in Hammer collection, Smithsonian Instution; *Washington Post*, March 25, 1907.

9 *Chicago Tribune*, April 24, 1910; *Washington Post*, December 26, 1897.

10 Denver in National Electric Light Association, *Report of the Committee on Progress* (1907); *Signs of the Times*, June 1916.
11 Cited in Leach, *Land of Desire*, 43–44.
12 Winston Churchill, *The Dwelling-Place of Light* (London: Macmillan, 1917), 16.
13 *Chicago Tribune*, January 3, 1893; *Anaconda Standard*, October 24, 1901; *Montgomery Advertiser*, October 31, 1901.
14 *Electrical Engineering,* September 27, 1893; Leonard de Vries and Ilonka van Amstel, *Victorian Inventions* (New York: American Heritage Press, 1972), 94; "Writing on the Clouds," *Youth's Companion*, July 26, 1894; *Illuminating Engineer* (May 1907), 229; *Electricity, a Popular Electrical Journal*, November 8, 1893.
15 William Dean Howells, *Impressions and Experiences* (1896), 270–71; Waldo Frank, *In the American Jungle (1925–1936)* (New York: Farrar & Rinehart, 1937), 117; E. A. Ross, *Changing America* (1910), 100–101.
16 *Christian Science Monitor*, March 15, 1909; *New York Times*, July 9, 1910; G. Glen Gould, "Where There Is No Vision," *Art World*, January 1917.
17 *Electrical World*, October 12, 1889; *American Gaslight Journal*, December 31, 1906; *Electrical Solicitor's Handbook* (1913); *Decorative and Sign Lighting Illustrated: A Report Made at the Twenty-Sixth Annual Convention of the National Electric Light Association Held at Chicago, Illinois, May 26, 27, 28, 1903* (1903).
18 *Printer's Ink*, May 13, 1903; *Chicago Tribune*, September 8, 1899; *Atlanta Constitution,* April 17, 1907.
19 David E. Nye, *American Technological Sublime* (Cambridge, MA: MIT Press, 1996), 188; on this conflict in New York, and the particular concern about billboards, see Michele Bogart, *Advertising, Artists, and the Borders of Art* (Chicago: University of Chicago Press, 1995), 90–105.
20 "Spectacular Lighting from the Esthetic Point of View," *Good Lighting* (1908), 162; Sir William Thomson cited in *Electrical World*, February 8, 1890.
21 Dietrich Neumann, *Architecture of the Night: The Illuminated Building* (London: Prestel, 2002), 42.
22 Gregory F. Gilmartin, *Shaping the City: New York and the Municipal Art Society* (New York: Clarkson Potter, 1995), 49.
23 Earlier attempts to develop "artistic" streetlamps are discussed in Rossell, "Compelling Vision," chapter 2; *The Nation*, May 30, 1901; Bulletin No. 7, Municipal Art Society of New York (1904), New-York Historical Society.
24 Gilmartin, *Shaping the City*, 55–59; *New York Times*, September 4, 1903; Michele H. Bogart, *The Politics of Urban Beauty: New York and Its Art Commission* (Chicago: University of Chicago Press, 2006), 12; *New York Times*, March 17, 1905.
25 The *Los Angeles Times* article is quoted at length in Feldman, *The Art of Street Lighting in Los Angeles* (Los Angeles: Dawson's Bookshop, 1972), 31–35.
26 *Lighting Journal*, August 1913.
27 "Civic Improvement from an Artistic Standpoint," *American Architect and Building News*, February 17, 1906; *New York Times*, February 19, 1913; Robinson, *Modern Civic Art,* 147; Frederic Howe, *European Cities at Work* (1913), 103; Glenn Marston, *The World To-Day* (1911), 296.
28 *Municipal Journal*, January 2, 1913, cited in Rossell, "Compelling Vision," 32; Rossell, "Compelling Vision," 40–45; Charles Mulford Robinson, *The Improvement of Towns and Cities: Or, the Practical Basis of Civic Aesthetics* (1906), 56–57.

29 *Municipal and County Engineering* (December 1921), cited in Rossell, "Compelling Vision," 38.

ELEVEN: ILLUMINATION SCIENCE

1 "The Illuminating Engineering Society, First Annual Meeting, Held in New York, January 14th," *Illuminating Engineer*, January 1, 1907; Edward Hyde, "The Physical Laboratory of the National Electric Lamp Association," *Abstract Bulletin*, December 1913.

2 Augustus D. Curtis, *The American Architect*, January 17, 1912; Chris Otter, *The Victorian Eye: A Political History of Light and Vision in Britain, 1800–1910* (Chicago: University of Chicago Press, 2008), 178.

3 Edward Rossell, "Compelling Vision: From Electric Light to Illuminating Engineering, 1880–1940" (PhD dissertation, University of California, Berkeley, 1998), xxvii; *Electrical Review*, 1907.

4 "Pseudo-Testimonials," *Illuminating Engineer*, May 1 and July 1, 1911; *Knoxville Sentinel*, March 18 and March 22, 1910; *Kansas City Star*, April 5, 1912.

5 Rossell, "Compelling Vision," 57; Laurent Godinez, *Display Window Lighting and the City Beautiful* (1914), 26.

6 Godinez, *Display Window Lighting*, 30.

7 On the creation of electric light displays, Stieringer declared these a "thoroughly modern creation. . . . We build on modern thought, modern intellect, modern knowledge; not on the way-behind conditions of centuries ago." Cited in Rossell, "Compelling Vision," 94, from "Electrical Installation and Decorative Work in Connection with the Exposition Buildings in the Pan-American Exposition," *Quarterly Bulletin of American Institute of Architects*, October 1901, 167. Sean Johnston, *A History of Light and Colour Measurement: Science in the Shadows* (Bristol, UK: Institute of Physics Publishing, 2001), 80–84.

8 *Electrical World*, May 17, 1890, August 2, 1890; Rossell, "Compelling Vision," 51–53; A. S. McCallister, "Illumination Units and Calculations," *Illuminating Engineering Practice* (1917), 1–35; the call to use watts instead of candlepower is in *Electrical World*, August 2, 1890. On the development of photometry prior to electric light, see Otter, *The Victorian Eye*, 154–68.

9 Rossell, "Compelling Vision," 54; Otter, *The Victorian Eye*, 168–72.

10 Arthur A. Bright Jr., *The Electric-Lamp Industry: Technological Change and Economic Development from 1800 to 1947* (New York: Macmillan, 1949), chapter 7.

11 *Illuminating Engineering Practice*, 57; E. L. Elliott, "Questions in Illuminating Engineering," *Illuminating Engineer* (1909), 168; *Electrical World*, February 8, 1890. For a review of the British counterpart to this discussion about vision and the dangers of improper lighting, see Otter, *The Victorian Eye*, chapter 1.

12 "Is Electric Light Injurious to the Eyes?," *Good Lighting and the Illuminating Engineer* (1908), 162; *Electrical World*, July 18, 1908, 118–19.

13 *Christian Advocate*, March 25, 1886; *Light: Its Use and Misuse, A Primer of Illumination* (1912), 2; Laurent Godinez, *Lux et Veritas* (1911); *Lighting from Concealed Sources* (1919), 22.

14 *Juice*, May 1911.

15 *Scientific American*, cited in *Crockery and Glass Journal*, May 6, 1915; Matthew Luckiesh, *Artificial Light: Its Influence upon Civilization* (1920), 303; E. S. Keene, *Mechanics of the Household* (1918), 306–13; Augustus D. Curtis, "Indirect Lighting," *Popular Mechanics*, January 1909; "Illuminate Tower of

Woolworth Building," *Popular Mechanics* (May 1915), 71; *New York Times*, April 25, 1913; on the integration of lighting into modern European architectural thought, see Werner Oechslin, "Light Architecture: A New Term's Genesis," in *Architecture of the Night: The Illuminated Building*, ed. Dietrich Neumann (London: Prestel, 2002); on the Woolworth Building and its precedents, see ibid., 55, 102.

16 On the contentious relationship between illumination engineers and architects, see "'Architecture of the Night' in the U.S.A," in Neumann, *Architecture of the Night*; "An English View of Illuminating Engineering," *Good Lighting and the Illuminating Engineer* (February 1908), 881; Laurent Godinez, "The Ultimate Relation of Illuminating Engineering to Public Utility Corporations," *Illuminating Engineer*, August 1, 1911.

17 For a good example of the use of bare bulbs in "handsomely decorated interiors," see *Electrical World*, October 4, 1890.

18 "Art Nouveau in Fixture Design," *Illuminating Engineer*, April 1, 1906; Alastair Duncan, *Art Nouveau and Art Deco Lighting* (New York: Thames & Hudson, 1978); Louis C. Tiffany, "The Tasteful Use of Light," *Scientific American*, April 15, 1911. By the 1920s, art deco designers more comfortably adapted to the new age of the machine rejected these lush decorative motifs, preferring clean geometric lines and white light that seemed better to reflect, rather than hide, the lamp's function. But they agreed with Tiffany that the bulb itself was best hidden from view, a goal more important than ever as lights grew brighter.

19 Luckiesh, *Artificial Light*, 312; Otter, *The Victorian Eye*, 61.

20 *The Hardware Review*, September 1917, 96; Leonard S. Marcus, *American Shop Windows*, 14–15; Glenn Marston, "American Public Lighting," *The World To-Day* (1911), 294; "Window Trimming," *The American Stationer*, March 26, 1910. On the development of store windows, see William Leach, *Land of Desire: Merchants, Power and the Rise of a New American Culture* (New York: Vintage, 1994), chapter 2.

21 Godinez, *Display Window Lighting*, 50.

22 Rossell, "Compelling Vision," 151; Leach, *Land of Desire*, 62–63, 76; Susan Porter Benson, "Palace of Consumption and Machine for Selling: The American Department Store, 1880–1940," *Radical History Review*, Fall 1979.

23 F. H. Bernard, "What Better Industrial Lighting Can Do to Stimulate Production," *Electrical Review*, September 6, 1919.

24 *Electrical Review*, September 6, 1919; Matthew Luckiesh, *Light and Work* (1924), 199; Daniel Nelson, *Managers and Workers: Origins of the New Factory System in the United States, 1880–1920* (Madison: University of Wisconsin Press, 1975), 29; Richard Gillespie, *Manufacturing Knowledge: A History of the Hawthorne Experiments* (Cambridge: Cambridge University Press, 1993).

25 *Nation's Health*, August 1921, 440–43; *Illuminating Engineer of National Lamp Works of General Electric Company Special Lecturer at Case School of Applied Science, Lecturer, Illuminating Engineering Course*; George Price, *The Modern Factory: Safety, Sanitation and Welfare* (1914), chapter 5.

26 "Education Hygiene," *Annual Report, Vol. 1, United States Office of Education* (1916), 329. "Over-use of the Eyes in Education," *Literary Digest*, September 24, 1910; Luckiesh, *Light and Work*, 284–85.

27 "Church Lighting," *Electrical Review*, November 1, 1913; F. Laurent Godinez, "Church Lighting," *The American Architect*, January 16, 1918; Kenneth Curtis, "Artificial Lighting in Church," *The American Architect*, December 31, 1924.

28 *Electricity*, February 20, 1895; Emile Perrot, "Church Lighting Requirements,"

Illuminating Engineering Practice, 297–305; Henry C. Horstmann, *Modern Illumination: Theory and Practice* (1912); *American Ecclesiastical Review* (February 1899), 206–7.

29 *The Illustrated American*, May 28, 1892.

30 S. J. Kleinberg, "Gendered Space: Housing, Privacy, and Domesticity in the Nineteenth-Century United States," in *Domestic Space: Reading the Nineteenth-Century Interior*, ed. Inga Bryden and Janet Floyd (Manchester: University of Manchester Press, 1999).

31 Sarah Milan, "Refracting the Gaselier: Understanding Victorian Responses to Domestic Gas Lighting," in Bryden and Floyd, *Domestic Space*; Otter, *The Victorian Eye*, 203.

32 *Better Electric Lighting in the Home: Bulletin, Engineering Department, National Lamp Works of General Electric Co.*, May 10, 1922. On the dangers of reading in bed, *The Medical Record*, January 2, 1904. In their guides to creating a properly electrified home, the English writers always offered a distinctively class-conscious twist, advising on the perennial question of "what to do about the servants?" While servant quarters were usually spartan, these manuals suggested that even the hired help should have incandescent light, though their rooms should be fitted with only the most utilitarian fixtures. Once the scullery maids and gardeners got over the superstitious fear of electrocution, they would come to love the light as much as their masters. Too much, perhaps. These guidebooks to a properly electrified home advised that a switch controlling the servants' lights should be placed in the master's bedroom. Since servants were "notorious wasters of light," as one put it, the master could flip the switch when he was ready for bed, plunging the entire house into darkness.

33 For the British discussion on the value of light to public health, see Otter, *The Victorian Eye*, 62–72; Mark Rose, *Cities of Light and Heat: Domesticating Gas and Electricity in Urban America* (University Park: Pennsylvania State University Press, 1995), 106–9.

34 Helen Campbell, *Household Economics: A Course of Lectures in the School of Economics of the University of Wisconsin* (1897); Peter C. Baldwin, *In the Watches of the Night: Life in the Nocturnal City, 1820–1930* (Chicago: University of Chicago Press, 2012), 163–64.

35 Mary Lockwood Matthews, *The House and Its Care* (1927).

36 Laurent Godinez, *The Lighting Book* (1913), 8.

37 Laurent Godinez, "The Ultimate Relation of Illuminating Engineering to Public Utility Corporation," *Illuminating Engineer*, August 1, 1911; Curtis, "Artificial Lighting in Churches," 31.

38 On the history of electrical blackouts, see David Nye, *When the Lights Went Out: A History of Blackouts in America* (Cambridge, MA: MIT Press, 2010).

TWELVE: RURAL LIGHT

1 David Nye, *Electrifying America: Social Meaning of a New Technology* (Cambridge, MA: MIT Press, 1992), 261; Chris Otter, *The Victorian Eye: A Political History of Light and Vision in Britain, 1800–1910* (Chicago: University of Chicago Press, 2008), chapter 5; *Harper's Weekly*, July 11, 1903; Arthur A. Bright Jr., *The Electric-Lamp Industry: Technological Change and Economic Development from 1800 to 1947* (New York: Macmillan, 1949), 212–13.

2 "Dangers of Electric Lighting," *Cincinnati Lancet and Clinic*, August 2, 1900; on bathtubs, see *New York Times*, July 16, 1907.

3 *Electrical World*, August 8, 1891; "Some Aspects of the Competition with Gas and Other Illuminants," *American Gaslight Journal*, January 29, 1894; Harold Platt, *The Electric City: Energy and the Growth of the Chicago Area, 1880–1930* (Chicago: University of Chicago Press, 1991), 153–54; *Scientific American*, April 5, 1913.

4 Nye, *Electrifying America*, 259–60.

5 Earl Anderson and O. F. Haas, "Illumination and Traffic Accidents: Statistics from Thirty-Two Cities," *Transactions of the Illuminating Engineering Society* (November 20, 1921), 453; Sidney Morse, *Household Discoveries* (1908), 94. Morse estimates that U.S. customers spent $11 million each year on acetylene, $60 million on "illuminating gas," $133 million on kerosene, and $150 million on electricity.

6 National Electric Light Association, *The Electrical Solicitor's Handbook* (1909), 33; "Business Development," *General Electric Review*, June 1922.

7 *Implement Age*, August 12, 1911; Nye, *Electrifying America*, 294. By 1927, one estimate suggested that "of the six and a half million farms in this country, less than three-quarters of a million farms have electric lighting." W. C. Brown, "Farm Lighting," *Bulletin, National Lamp Works of General Electric Co.*, September 15, 1927.

8 Helen Campbell, *The Easiest Way In Housekeeping and Cooking* (1903), 48–50; Morse, *Household Discoveries*, 104.

9 Morse, *Household Discoveries*, 98.

10 *Electrical Engineering*, December 1914.

11 Susan Prendergast Schoelwer, "Curious Relics and Quaint Scenes: The Colonial Revival at Chicago's Great Fair," in *Colonial Revival in America*, ed. Alan Axelrod (New York: Norton, 1985), 204.

12 Clarence Cook, *The House Beautiful* (1878), 123, 273; Morse, *Household Discoveries*, 106–9; Catherine Beecher, *The American Woman's Home* (1869), 362; *House and Garden*, March 1921.

13 In the early twentieth century, most American restaurants featured bright light, patrons being evidently eager "to see or be seen throughout the room." The French, by contrast, preferred the intimacy and mood of candlelight. *Illuminating Engineering Practice*, 264.

14 Richard Spillane, "The Moles of New York," *Forbes*, January 6, 1923.

15 G. Stanley Hall, "Reactions to Light and Darkness," *American Journal of Psychology*, January 1903. On Hall's ideas about human development and civilization, see Gail Bederman, *Manliness and Civilization: A Cultural History of Gender and Race in the United States, 1880–1917* (Chicago: University of Chicago Press, 1995).

16 Dr. A. T. Bristow, "The Most Healthful Vacation," *The World's Work*, June 1903; Sarah Burns, *Inventing the Modern Artist: Art and Culture in Gilded Age America* (New Haven: Yale University Press, 1999), 209; Horace Kephardt, *Camping and Woodcraft* (1919), 19–20.

17 *New York Tribune*, August 26, 1916; H. S. Firestone, "My Vacations with Ford and Edison," *System: The Magazine of Business*, May 1926.

18 *Wilkes-Barre Times Leader*, August 21, 1918; *Inquirer*, September 9, 1918; *San Francisco Chronicle*, July 24 and August 6, 1921.

19 "The Value of Proper Illumination," *Journal of Electricity*, February 15, 1920, cited in Edward Rossell, "Compelling Vision: From Electric Light to Illuminating Engineering, 1880–1940" (PhD dissertation, University of California, Berkeley, 1998), 72; on Western disillusionment with technology and invention in the

aftermath of World War I, see Michael Adas, *Machines as the Measure of Men: Science, Technology, and Ideologies of Western Dominance* (Ithaca: Cornell University Press, 1989), chapter 6.

20 Thomas Edison, "What Is Life?," *Cosmopolitan*, May 1920.

21 *New York Times*, March 18, 1930.

22 *The Home of a Hundred Comforts*, Merchandise Department, General Electric Corporation (1920); Henry Schroeder, "History of the Electric Light," *Smithsonian Misc. Collections*, vol. 76 (1923), 94; Nye, *Electrifying America*, 296–99.

23 "Electricity to End Farm Drudgery," *Popular Mechanics*, August 1925; Morris L. Cooke, "REA—New Light to the Farm," *Independent Woman*, January 1936; *Saturday Evening Post*, March 19, 1938.

24 "Electricity on the Farm," *Science*, July 4, 1930; "Effect of Rural Electrification upon Farm Life," *Monthly Labor Review*, April 1939; *Time*, July 4, 1938; *Christian Science Monitor*, July 18, 1935; Nye, *Electrifying America*, 325.

25 "Will Villages Vanish?," *Christian Science Monitor*, May 20, 1930; Nye, *Electrifying America*, 291.

26 On New Deal electrification policy, see Ronald C. Tobey, *Technology as Freedom: The New Deal and the Electrical Modernization of the American Home* (Berkeley and Los Angeles: University of California Press, 1996), chapter 4.

27 Franklin Delano Roosevelt cited in Nye, *Electrifying America*, 304.

EPILOGUE: ELECTRIC LIGHT'S GOLDEN JUBILEE

1 Jill Jonnes, *Empires of Light: Edison, Tesla, Westinghouse, and the Race to Electrify the World* (New York: Random House, 2003), 347–53.

2 *Light*, February 1927; *New York American*, October 19, 1931; Matthew Josephson, *Edison: A Biography* (New York: McGraw-Hill, 1959), 473.

3 *Los Angeles Times*, October 30, 1929; Thomas Hughes, *American Genesis: A Century of Invention and Technological Enthusiasm* (Chicago: University of Chicago Press, 1989), 169.

4 J. W. Milford, "The Second Fifty Years: An Interview with Thomas A. Edison," *Revenue*, January 19, 1930.

5 Ibid.

Illustration Credits

Frontispiece: Thomas Alva Edison photographed in his laboratory by George Grantham Bain, February 1908. Culver Pictures/The Art Archive at Art Resource, NY.
Page 11: *Electrical World*, 12 April 1884. *La Lumiere Electrique* (1881).
Page 16: The Huntington Library, San Marino, California.
Page 22: *La Lumiere Electrique* (1881).
Page 26: *Harper's Weekly*, 14 January 1882.
Page 37: Algave and Bouldard, "The Electric light, its history, production, and applications," 1884.
Page 39: *La Lumiere Electrique* (1881).
Page 46: *La Lumiere Electrique* (1881).
Page 50: *La Lumiere Electrique* (1881).
Page 54: *The Electrical Review*, New York (March 7, 1885).
Page 64: San Jose tower, December 1881, image via Wikimedia.
Page 75: U.S. Dept. of the Interior, National Park Service, Thomas Edison National Historical Park.
Page 77: From *Puck*, Oct. 23, 1878. Library of Congress, call number AP101.P7 1878 (Case X).
Page 82: *Harper's Weekly*, 14 May 1881.
Page 88: *Electrical World*, 21 June 1890. The Huntington Library, San Marino, California.
Page 95: The Huntington Library, San Marino, California.
Page 101: *Harper's Weekly*, 3 March 1883.
Page 107: *Electrical World*, *18 July* 1891.
Page 112: Collection of The New-York Historical Society. Costume Ball Photograph Collection, negative #39500.
Page 115: *Electrical World*, 28 June 1890.
Page 121: Detroit Publishing Co., Library of Congress, call number LC-D4-10727.
Page 123: *The Graphic*, 13 August 1904.
Page 125: *Harper's Weekly*, 19 September 1891. The Huntington Library, San Marino, California.
Page 130: *Scientific American*, 15 November 1884.
Page 133: *Electrical World*, 11 October 1890. The Huntington Library, San Marino, California.
Page 137: Institute of Electrical and Electronics Engineers History Center Archives.
Page 142: Lester S. Levy Collection of Sheet Music, Special Collections, Sheridan Libraries, Johns Hopkins University.

Page 149: *Harper's Weekly*, 11 April 1891. The Huntington Library, San Marino, California.

Page 156: *Electrical World, 12 February 1887.*

Page 165: *Electrical World*, 19 July 1884. The Huntington Library, San Marino, California.

Page 166: *Transactions*, Illumination Engineering Society, Vol. 10 (1915).

Page 170: *La nature* 1268 (11 September 1897), p. 225.

Page 172: John Harvey Kellogg, *Rational Hydrotherapy: A Manual of the Physiological and Therapeutic Effects of Hydriatic Procedures, and the Technique of their Application in the Treatment of Disease* (Philadelphia: F. A. Davis, 1903).

Page 178: *Electrical World*, Feb. 28, 1891. The Huntington Library, San Marino, California.

Page 183: *Judge*, Vol. 17, No. 419, 26 October 1889, cover.

Page 188: *Leslie's Weekly.*

Page 189: *Electrical World*, 1 March 1890.

Page 192: *Electrical World*, January 1891. The Huntington Library, San Marino, California.

Page 211: *Electrical World*, 9 January 1886. The Huntington Library, San Marino, California.

Page 213: *Electrical World*, 15 February 1890. The Huntington Library, San Marino, California.

Page 221: Forbes Co., Boston and New York, circa 1883. Library of Congress, call number POS-TH-KIR, no. 20 (C size).

Page 228: Dime Novel Collection, Special Collections Department, Tampa Library, University of South Florida, Tampa, Florida.

Page 232: Charles Dudley Arnold, photographer. Library of Congress, call number LOT 4654-2.

Page 239: Library of Congress, call number LC-USZ62-125033, U.S. GEOG FILE—New York—New York City—Buildings.

Page 244: Collection of The New-York Historical Society. Irving Browning, negative #58362.

Page 247: *Transactions*, Illuminating Engineering Society, Vol. 8 (1913), p. 616.

Page 248: Arthur Williams, "Decorative and Sign Lighting Illustrated: A Report Made at the Twenty-Sixth Annual Convention of the National Electric Light Association, New York, 1903. The Huntington Library, San Marino, California.

Page 255: *The Municipal Art Society* pamphlet, 1903.

Page 256: *Night in Los Angeles* (1912). The Huntington Library, San Marino, California.

Page 258: Library of Congress, call number LC-USZ62-116320, U.S. GEOG FILE—Illinois—Chicago—Street views—State Street.

Page 259: *Electrical World*, 12 December 1891.

Page 265: General Electric pamphlet, 1907.

Page 269: Detroit Publishing Co. Collection, Library of Congress, call number LC-D4-73062.

Page 273: Peoples Drug Store, 7th & K, [Washington, D.C.], night. Library of Congress, call number LC-F82-4062.

Page 289: *Home of a Hundred Comforts*, General Electric pamphlet (1925).

Page 296: From the Collections of the Henry Ford, ID number THF25119.

Page 300: Lester Beall, artist. Rural Electrification Administration, U.S. Department of Agriculture. Library of Congress, call number POS-US.B415, no. 9.

Page 307: U.S. Dept of the Interior, National Park Service, Thomas Edison National Historical Park.

Index

Pages with illustrations are in italics.